PERGAMON INTERNATIONAL LIBRARY
of Science, Technology, Engineering and Social Studies
The 1000-volume original paperback library in aid of education, industrial training and the enjoyment of leisure
Publisher: Robert Maxwell, M.C.

AGRICULTURAL STATISTICS
A Handbook for Developing Countries

THE PERGAMON TEXTBOOK
INSPECTION COPY SERVICE

An inspection copy of any book published in the Pergamon International Library will gladly be sent to academic staff without obligation for their consideration for course adoption or recommendation. Copies may be retained for a period of 60 days from receipt and returned if not suitable. When a particular title is adopted or recommended for adoption for class use and the recommendation results in a sale of 12 or more copies, the inspection copy may be retained with our compliments. The Publishers will be pleased to receive suggestions for revised editions and new titles to be published in this important International Library.

Other Titles of Interest:

BALASSA, B.
Policy Reform in Developing Countries

COLE, S.
Global Models and the International Economic Order

CLARKE, J.I.
Population Geography and the Developing Countries

ECKHOLM, E. P.
Losing Ground: Environmental Stress and World
Food Prospects

JOLLY, R.
Disarmament and World Development

KOENIGSBERGER, O. H. & GROAK, S.
Construction and Economic Development:
Planning of Human Settlements

MENON, B. P.
Global Dialogue: The New International Economic
Order

OPENSHAW, K.
Cost and Financial Accounting in Forestry:
A Practical Manual

SINHA, R. & DRABEK, A.
The World Food Problem: Consensus and Conflict

TICKELL, C.
Climatic Change and World Affairs

AGRICULTURAL STATISTICS
A Handbook for Developing Countries

by

N. M. IDAIKKADAR
M.A.(Cantab), M.Sc.(London)

Former Statistician of the Food and Agriculture Organization of the United Nations

PERGAMON PRESS
OXFORD · NEW YORK · TORONTO · SYDNEY · PARIS · FRANKFURT

U.K.	Pergamon Press Ltd., Headington Hill Hall, Oxford OX3 0BW, England
U.S.A.	Pergamon Press Inc., Maxwell House, Fairview Park, Elmsford, New York 10523, U.S.A.
CANADA	Pergamon of Canada, Suite 104, 150 Consumers Road, Willowdale, Ontario M2J 1P9, Canada
AUSTRALIA	Pergamon Press (Aust.) Pty. Ltd., P.O. Box 544, Potts Point, N.S.W. 2011, Australia
FRANCE	Pergamon Press SARL, 24 rue des Ecoles, 75240 Paris, Cedex 05, France
FEDERAL REPUBLIC OF GERMANY	Pergamon Press GmbH, 6242, Kronberg-Taunus, Pferdstrasse, Federal Republic of Germany

Copyright © 1979 N. M. Idaikkadar

All Rights Reserved. No part of this publication may be reproduced, stored in a retrieval system or transmitted in any form or by any means: electronic, electrostatic, magnetic tape, mechanical, photocopying, recording or otherwise, without permission in writing from the publishers

First edition 1979

British Library Cataloguing in Publication Data
Idaikkadar, N. M.
Agricultural statistics – (Pergamon International Library).
1. Underdeveloped areas – Agricultural industries – Statistics
I. Title
338.1'01'82   HD1417   78-41193

ISBN 0-08-023388-0 (Hardcover)
ISBN 0-08-023387-2 (Flexicover)

Reproduced, printed and bound in Great Britain by Fakenham Press Limited, Fakenham, Norfolk

DEDICATED
TO THE
TWO GREAT WOMEN OF MY LIFE
MY MOTHER AND MY WIFE

Contents

    Preface   xi

1. **Introduction**   1
   1.1 *Background*   1
   1.2 *Definition*   1
   1.3 *Basic and current statistics*   2
   1.4 *Development of statistics*   3
   1.5 *Treatment of the subject*   4

2. **Methodology of Development**   6
   2.1 *Scope*   6
   2.2 *Stages of development*   6
   2.3 *Objective methods and use of sampling*   10
   2.4 *Objective methods and pilot surveys*   12
   2.5 *Gaps in data collection*   13

3. **Agricultural Production – General Ideas**   14
   3.1 *Importance*   14
   3.2 *Crops*   14
   3.3 *Direct estimation*   16

4. **Agricultural Production – Crop Yields**   19
   4.1 *Introduction*   19
   4.2 *Crop yield surveys*   19
   4.3 *Example (temporary crop)*   28
   4.4 *Example (permanent crop)*   31

5. **Agricultural Production (Crops) – Area**   33
   5.1 *Introduction*   33
   5.2 *Reporting system*   33

	5.3 Guidelines for development	35
	5.4 Area measurements	40
	5.5 Priorities	42
	5.6 Mixed cropping	42
	5.7 Permanent crops	44
6.	**Agricultural Production (Crops) — Forecasting**	46
	6.1 Need	46
	6.2 The two components	46
	6.3 Crop yields	47
7.	**Agricultural Production — Livestock and Livestock Products**	53
	7.1 Introduction	53
	7.2 Livestock inventory	53
	7.3 Livestock products	60
8.	**Index Numbers of Agricultural Production**	69
	8.1 General remarks	69
	8.2 Concept of production	69
	8.3 Weights	70
	8.4 Formula used	70
	8.5 Indices used	71
	8.6 Example	72
9.	**Supply/Utilization accounts (Food Balance Sheets)**	74
	9.1 Introduction	74
	9.2 Purpose and use	74
	9.3 Commodity coverage	75
	9.4 Concepts and definitions	77
	9.5 Guidance on filling the balance sheets	80
	9.6 Example	81
10.	**Census of Agriculture**	84
	10.1 General remarks	84
	10.2 Objectives	85
	10.3 Scope and coverage	86
	10.4 Concepts and definitions	88
	10.5 Planning and procedure	90
	10.6 Analysis, tabulation and final report	93

11. Survey and Sampling Methods — 96
- 11.1 *Scope* — 96
- 11.2 *Surveys* — 96
- 11.3 *Sampling methods* — 97

12. Price Statistics — 105
- 12.1 *Importance* — 105
- 12.2 *Some characteristics* — 105
- 12.3 *Types of agricultural prices* — 107
- 12.4 *Collection* — 109
- 12.5 *Users* — 112

13. Statistics for Agricultural Planning — 113
- 13.1 *Inclusion in existing data collection* — 113
- 13.2 *Land utilization* — 113
- 13.3 *Irrigation* — 116
- 13.4 *Fertilizer* — 117
- 13.5 *Employment in agriculture* — 117
- 13.6 *Agricultural power and machinery* — 118
- 13.7 *Agricultural credit* — 118
- 13.8 *Market intelligence* — 119
- 13.9 *Cost of production* — 120

14. Staff and Organization — 122
- 14.1 *Introduction* — 122
- 14.2 *Staff* — 122
- 14.3 *Selection and training* — 123
- 14.4 *Staff management* — 126
- 14.5 *Field organization* — 127

Appendices

Typical form for data collection — cost of production — 132
Typical form for data collection — reporting system

Relevant Publications — 134

Index — 137

Preface

This book is based on a series of lectures given on development of agricultural statistics in some countries of Asia and Africa where the author had served as Statistician to the Food and Agriculture Organization of the United Nations; but it took final shape only when the author gave a course of lectures on Third World Agricultural Statistics and Survey and Sampling Methods to M.Sc. students of agricultural economics at the University of Oxford in 1976.

The method of treatment and the arrangement of topics are intended to build up a coherent sub-discipline within the framework of statistics, out of the various sections of the subject, which so far have been dealt with piecemeal. An endeavour has been made to cover the entire field of agricultural statistics.

The aims and scope of the book are:
  (a) To provide guidelines to those in charge of agricultural statistics in developing countries, for improving their statistics in a systematic way, to know their priorities and to have clear objectives;
  (b) To provide for planners, policy makers and senior statisticians a handbook on the methodology of agricultural statistics;
  (c) To emphasize the use of objective methods for collection of data;
  (d) To emphasize the importance of collecting data independent of farmers to overcome their subjective attitude to data collection;
  (e) To use sampling with its accompanying benefits, in data collection.

The book has been written in the first instance for use in Statistics Departments, Ministries of Agriculture and Planning Departments of the developing countries. Senior officers in these Departments would be most interested in the methodology while the junior officers may prefer the practical approach. With his experience in giving lectures on the subject in five Universities in the developing countries and one in a developed country, the author feels the book may be of some use to students of statistics, economics and agriculture in the Universities in both developing and developed countries.

The author is grateful to the late Mr. K. E. Hunt, Director of the Institute of Agricultural Economics, University of Oxford, for providing facilities at the Institute for writing the book and for the valuable suggestions made at every stage of its progress.

The stimulus for writing the book would not have been possible but for the experience gathered by the author in the Food and Agriculture Organization of the United Nations. The technical reports released by the FAO from time to time have also helped him.

For brevity, sometimes the term "statistics" is used instead of "agricultural statistics" in the text.

N. M. Idaikkadar
1978

CHAPTER 1

Introduction

1.1 Background

When introducing the subject of agricultural statistics for developing countries, one has to appreciate the position of agriculture in these countries.

Agriculture plays a very important part in their economic life and is likely to do so for many decades. The contribution of agriculture to the gross domestic product generally exceeds 50%. A large portion of the population is sustained by agriculture. Further, agriculture is an important foreign exchange earner and could also save foreign exchange by producing locally those agricultural products now imported into the country.

Due to the important position agriculture holds in the life of these countries, promotion of agriculture is a key function of the Governments concerned. For success in promoting agriculture, planning is essential. No planning is meaningful unless backed by factual information. Thus statistical data are required for any agricultural planning. Further, statistical data on agriculture are needed for the life of the community in assessing the requirements for feeding the people, quantities available for export and import and various other purposes.

Thus, regular collection of agricultural statistics is essential to the functioning of the community and for its agriculture and agricultural development.

1.2 Definition

It is useful for us to be clear as to what we mean by agricultural statistics. We may define agricultural statistics as the aggregate of numerical information of different fields of agriculture and its economy. The different fields of agriculture would embrace food crops, commercial crops, livestock and livestock products and the different fields of agricultural economy would cover all that is modern agricultural economics.

1.3 Basic and Current Statistics

Agricultural statistics can be divided into two broad groups: (a) basic, and (b) current.

Basic statistics are those which deal with the enduring characteristics of agriculture, such as land utilization, land tenure, distribution of holdings etc. Such statistics are collected by countries in their decennial agricultural censuses. In basic statistics, estimates are made for individual units (like holdings).[1]

Current statistics provide information on agricultural activities performed more or less continuously every year, e.g. the area and production of temporary and permanent crops, production of meat, milk and eggs. Such statistics are collected annually or at more frequent intervals. In current statistics we are generally interested in national or provincial totals or averages.[2]

In this book we shall mostly be dealing with current statistics.

Besides current and basic statistics, a new requirement has come in, not covered by census and current statistics. Such statistics are not required annually but possibly cannot wait for a census every ten years. These deal with economic aspects of agricultural holdings not normally covered by census or current statistics. In these cases data are collected by surveys conducted as and when needed, perhaps once in two or three years. In the Food and Agriculture Organization of the United Nations such surveys are termed National Farm Surveys or General Economic Surveys of Agriculture.

---

[1] Such statistics are called micro-statistics.

[2] Such statistics are called macro-statistics.

1.4 Development of Statistics

For promotion of planning in agriculture and for other needs of agricultural data, many developing countries are making efforts to develop their agricultural statistics. Historically these statistics were gathered as by-product of data collected for administrative purposes by the Government Departments concerned[1] with the inherent defect that such collections are made by subjective methods and by administrative officers not too conversant with basic statistical techniques. Over the years National Statistical Offices emerged in these countries and started collecting the data directly or with the assistance of the Ministry/Department of Agriculture. With the advent of National Statistical Offices, efforts have been and are being made in improving the quality of data — by improving the reporting system and gradually replacing some of the subjective methods of collection by objective methods using sampling techniques. Progress however has varied from country to country depending on a number of factors as described in the succeeding paragraph.

In the process of development of statistics in a country one has to face the approach of the policy makers, planners and statisticians to the subject. It has been the author's experience that policy makers who are in charge of agricultural statistics in the developing world have little knowledge of what the discipline of agricultural statistics is or what its development means. Some policy makers and planners have vague ideas on its development and others though knowledgeable have rarely addressed their minds to the subject. Further the statistician who handles the subject at the national level and who could guide others has his hands full with routine statistics and rarely addresses his mind to its overall development — he generally handles the subject piecemeal depending on individual tastes and *ad hoc* requirements of the Government. There is thus a real need for some guidelines to those concerned on the subject for its systematic development, which the author attempts to describe in this book. The general trend in the guidelines would be the gradual replacement of some of the existing subjective methods where possible and the greater use of objective methods.

---

[1] Such gathering of data is called the reporting system.

It may be asked why stress is placed on the replacement of some of the subjective methods by objective methods. Besides the normal merits of the objective methods over subjective methods, the latter give rise to very serious errors especially in developing countries. Many farmers are smallholders and they lack sufficient education and knowledge to report on their agriculture, making it necessary for the State to use its officers or to specially employ enumerators for collecting the needed data. So, to have dependable statistics, it is necessary to use methods and techniques which are independent of the farmers themselves. Such techniques might be direct observations by specially selected personnel for that purpose and physical measurements.

In the use of objective methods, sampling is indispensable. Sampling techniques are used in agricultural statistics for the same reasons as in the other fields of statistics. Namely, by reducing the volume of the material to be dealt with, we reduce the cost of work, and make it possible to improve the quality of data because of the reduced number and better selection of personnel.

Gradual replacement of some of the subjective methods by objective methods and use of sampling techniques, have to be fitted into the existing scheme of collection of agricultural data of the country gradually. No grandiose expansion of the statistical staff should be requested all at once from a Government nor a request for large funds. The process of development of statistics has to be gradual and acceptable to the Government concerned. Priorities have to be made in the use of new methods and techniques and phased over time taking into consideration the available staff and funds.

It is not always possible to replace some of the subjective methods used in agricultural statistics by objective methods as some enquiries do not lend themselves to objective measurements in the field; e.g. data on prices and some items on cost of production and food consumption surveys. In such cases other techniques will have to be used.

1.5 Treatment of the Subject

The chapters that follow start with methodology of development. This is illustrated by its application in turn to current statistics and basic statistics. Thereafter other agricultural statistics are dealt with.

In current statistics, agricultural production is dealt with in great detail. This involves production of temporary and permanent crops as well as that of livestock and livestock products. In estimating crop production, the two components area cropped and yield rate for temporary crops and number of (bearing) trees and yield per (bearing) tree for permanent crops are described. Some ideas on forecasting of crop production are given as well as on index of agricultural and food production.

The important subject of food supply/utilization accounts (food balance sheets) is dealt with followed by a description of a census of agriculture.

There is one chapter on surveys and sampling methods summarizing in one place all that has been dealt with on the subject in the whole book.

Statistics on land utilization, irrigation and fertilizer as well as on prices, market intelligence and cost of production are described under statistics for agricultural planning.

The concluding chapter is on staff and organization.

CHAPTER 2

Methodology of Development

2.1 Scope

The development of statistics may be considered as consisting of:
 (a) Improvement of the methods of collection of existing statistics for greater dependability;
 (b) Collection of statistics in areas where there are none; in other words filling the gaps in the existing data collection.

Let us first consider improvement of existing statistics.

2.2 Stages of Development

Countries in the developing world are in different stages of improving their statistics. Some countries are still using mainly subjective methods while others are using a combination of subjective and objective methods, the distribution of the two methods depending on the statistical environment of the country concerned. There is a third group which is mainly using objective methods; this group, however, is very small indeed.

We could thus broadly think of development in three stages:
 Stage I: Mainly using subjective methods of collection;
 Stage II: Using subjective methods of collection of some data, and objective methods for others;
 Stage III: Mainly using objective methods.

It is obvious that any "stage" of collection mentioned above, would provide better quality data than the preceding "stage". The ideal would be Stage III. This, however, would take a considerable period of time for a developing country to attain.

2.2.1 Stage I

Let us take a country in Stage I of development. If the country is going to take some time to get to Stage II due to limitations of finance and trained manpower it should not remain idle in the meantime. Improvements are possible, though limited due to the very nature of collection of statistics being subjective. The unit of enumeration plays an important part in the quality of data collected in production statistics, the smaller the unit the better the quality. Some improvement could be effected by having as small a unit as possible. For example, in estimating cropped area there are cases where villages are enumeration units; in cases like these, it is certainly an improvement if the unit of enumeration were reduced to a farm or a field. Another factor that affects the estimation in subjective methods is the type of person who estimates for a reporting unit. In many cases in developing countries such persons are village officers with little or no statistical knowledge. In these cases improvement in the quality of statistics could be effected by training these officers and periodically supervising their work by statistically trained staff for better collection of data. Supervision can take the form of selecting a random sample and checking on the reporting made. If the sample is large enough, suitable correction factors can be evolved for correcting the estimates.

2.2.2 Stage II

We may next consider a country in Stage II. The country has emerged from Stage I and is attempting to replace gradually some of its subjective methods by objective ones. It is useful in such a case to list all the subjective methods used and a study made to give priorities as to which existing methods should be replaced by objective methods earlier and which later; in other words, if the list has 10 subjective methods, these ten should be arranged in order of priority for adopting objective methods. A working rule would be to give top priority for replacement of existing subjective methods that give the most errors and for statistics where the country's economy is most concerned. Let us take a few examples:

(a) In estimating crop production, priority is given to the yield rate component by objective methods over the cropped area component, reason being that in estimating yield rate

subjectively one makes a greater error (percentagewise) than in estimating area under crop. (The error in area is very small where the land is cadastrally[1] surveyed.)
 (b) In estimating the cropped area in a country which has two distinct sectors (organized and non-organized), priority is given to the non-organized sector over the organized sector for introducing objective methods as the error in existing methods of estimation of non-organized sector would be much more than that for the organized sector. (In one country it was found that the error rarely exceeded 5% for the organized sector and about 30% for the other.)
 (c) If in a country, production estimates are made by subjective methods for two crops, and funds and staff are available for only one crop to be estimated by objective methods, the more important of the two crops has to be taken.

The priorities should be determined in consultation with the users of agricultural statistics, planners and policy makers. A programme of work should be drawn and the introduction of objective methods should be made systematically over the years. When such a programme is available in a Central Statistics Office/Ministry of Agriculture, there is every chance of continuity of the work and no one will lose sight of the objectivity. The author has had experience in some countries where adviser after adviser had given "pieces of advice" without dovetailing into a proper programme and in such cases there had been little progress. When there is a change in staff, the new person or persons can proceed from where the predecessors in office had left, when there is a programme. Otherwise there is always the danger that different officers would cover the same ground with the consequence of wastage of funds which the countries can ill afford and delays in the development of statistics.

2.2.3 Stage III

Let us now consider the countries in Stage III. There are not more than five countries at this stage in the developing world. They have come to this

---

[1] A cadastral map is a plan drawn to a sufficiently large scale to show property boundaries and individual buildings, and is used primarily for fiscal purposes.

stage as the Governments in these countries were the first to realize the value of dependable statistics. With their own strength and with the aid of international agencies these countries quickly switched over to the use of objective methods where possible. The only advice that can be given to these countries is that they should maintain what they have achieved and proceed further with more sophisticated methods.

2.2.4 Selection of priorities

An illustration is given here as to how priorities are made in replacing subjective methods by objective ones in Stage II in production statistics. Let us take a country and consider the development of statistics to crops. This country has four main food crops (rice, maize, cassava, soya-bean) and some minor food crops; there are also two cash crops, tea and rubber. We are interested in drawing a programme of development of current statistics for this country. In current statistics we are mostly concerned in production and its components, area and yield rate. Let us examine the present state of agricultural statistics in the country, and present it in a tabular form showing the methods prevailing:

Crops	Methods of Collection		
	Area	Yield rate	Production
Main Food Crops	Subjective by fields (for cadastral areas) by villages (for non-cadastral areas)	Objective (by sub-districts)	–
Minor Food Crops	Subjective (by villages)	Subjective (by sub-districts)	–
	Standing and Harvested Area	Yield Rate	Production
Cash Crops	Subjective (by estate superintendents for large estates and by sub-districts for small holdings)	–	Objective (by factories or processing centres)

(It is not necessary to estimate all three — area, yield rate and production — if we know any two, the third can be calculated.)

10   *Agricultural Statistics: A Handbook for Developing Countries*

This form shows that objective methods have already been adopted for the main food crops for determining yield rates. These consist of crop cuts using random sampling techniques and physical measurement of the crop in the selected plots. The form also shows that objective methods are used for cash crops for estimating the production. These estimates are compiled from accurate records maintained at the several factories of green tea leaf received and black tea produced and raw rubber received and dry rubber produced (total production of the country finds its way to the factory both from large estates and small holdings). It may be mentioned that in this country large estates dominate the tea and small holdings dominate the rubber.

On the principles described in Section 2.2.2, priorities in the following order are given for the use of objective methods:

(i) Area (standing and harvested) for rubber small holdings;
(ii) Yield rate for minor food crops;
(iii) Area (standing and harvested) for tea small holdings;
(iv) Area (non-cadastral) for main food crops;
(v) Area (non-cadastral) for minor food crops;
(vi) Area (cadastral) for main food crops;
(vii) Area (standing and harvested) for large tea estates;
(viii) Area (standing and harvested) for large rubber estates;
(ix) Area (cadastral) for minor food crops.

Priorities given here cannot be said as the last word on the subject. The list gives an indication of how priorities can be determined by taking yield rate before area, major crop before minor crop, and non-cadastral before cadastral. It is debatable whether cash crops should be given preference over food crops or vice versa — it would depend on the importance of the crop and how the country views it.

The country can now draw up a programme of work for development on the priorities given.

2.3   Objective Methods and Use of Sampling[1]

When adopting objective methods in statistics, one must face the reality of finding the large staff and sufficient funds for carrying them out. It is

---

[1] For a fuller description of sampling see Chapter 11. Those not very familiar with sampling would be advised to read that chapter before reading this section.

not easy to find the needed money in developing countries for statistics as previously stated. Even when funds are made available, it is not easy to find the needed staff. So we resort to sampling in our objective method approach for:

(i) Reducing the cost;
(ii) Reducing the staff;
(iii) Reducing the volume of material to be dealt with;
(iv) Improving the quality of data (because of the reduced number and better selection of personnel).

Objective methods involve physical measurement of the characteristic — in determining yield rates crop cuts are done, in area surveys measurements and in fruit statistics counting.

In agricultural statistics, application of sampling with objective methods is found mainly in:

(i) Yield surveys of temporary and permanent crops;
(ii) Yield surveys for horticulture;
(iii) Area measurements.

Extensive work has been done in India and some other countries in determining yield rates of temporary crops by using sampling techniques and physical measurement of the crops. In this method small plots are selected in cultivators' fields by random sampling techniques and the crop is harvested and weighed by enumerators trained for this work. This is done for the whole country and yield rates with sampling errors are made available at national and lower levels.

Crop cut on permanent crops has not been done as extensively as on temporary crops. It is partly due to estimates of production being readily available for commodities like tea, rubber and cotton from the factories that there is no need for direct determining of yield rates. The other reason being that the work is far more arduous than on temporary crops. Preparing the frame, separating the bearing palms from the non-bearing palms, drawing a sample of cluster of bearing palms and harvesting periodically for long periods (compared to harvesting in a few hours the sample plot for temporary crops) make this work sometimes out of reach for some countries. For permanent crops, cluster (of trees) takes the place of plot in temporary crops. Countries could adopt objective methods for determining yield rates for permanent crops wherever direct estimation of production is not possible.

In horticulture the pattern of work for fruit trees would be similar to that for permanent crops and for vegetables like that of temporary crops. Countries have been slow in adopting objective methods for horticulture perhaps due to less importance being attached to it compared to food crops (cereals, starchy roots and pulses) and cash crops (tea, rubber, coffee). However, objective methods could be used for this category with profit as in the other cases.

In area measurements, one starts with non-cadastral areas and then proceeds to cadastral ones for reasons stated earlier. The sampling unit could be a holding (if not large) or a parcel or a field. The ideal measurement would be one by proper land surveyors. But sufficient funds and land surveyors are generally lacking and statistical staff resort to other ways of measurements not so accurate as land surveyors'. Simple instruments are available for measuring lengths and angles and enumerators can be trained for this work as has been done by some countries which have succeeded in area measurements on a sampling basis. Experience has shown that for area measurements, measurements of lengths by enumerators give far more satisfactory results than angles. Hence when measuring a holding, parcel or field, it will be preferable to split them into triangular plots and measure the three sides to get the area. With the experience gathered in the few countries where area measurements have been conducted, countries with large portions of land not cadastrally surveyed could make a start with this method even on 5% of their lands annually so that in about 20 years, the whole country would be covered.

Sampling has also been applied to other fields in agricultural statistics. These statistics include livestock, livestock products, food consumption surveys, checking quality of statistical data, censuses and agricultural experimentation. It has not been found possible to apply entirely objective methods in these cases due to practical problems in the field.

2.4 Objective Methods and Pilot Surveys

In all statistical work pilot surveys are essential before embarking on full-fledged surveys. These are needed to obtain the facts to build a

Methodology of Development

rational survey design.

When replacing a subjective method by an objective one using sampling, the first step is to make some pilot surveys with the method proposed. These surveys would reveal both quantitative and qualitative information — quantitative in giving estimates of the components of variation, averages, percentages, coefficients of correlation, the funds or time needed to carry out an operation etc. — qualitative regarding definitions, questionnaires, field difficulties, etc. Sometimes more than one pilot survey is needed to sort out these problems.

If, for instance, it is proposed to have crop cuts to determine the yield rates objectively, it is useful to conduct some pilot surveys on one or two crops in small areas like sub-districts or groups of villages. These surveys would indicate the efficiency of the sample design, the cost-estimate, field difficulties (like getting at the farmer at the right time, or in harvesting or weighing the produce) and other problems connected with translating into the field the methods finalized in the office. With the experience gathered in the pilot studies, various aspects of the main survey could be modified for greater efficiency.

2.5 Gaps in Data Collection

It was mentioned in 2.1 (b) that in the process of development, we have to collect statistics in areas where there are none. Gaps in statistics come to light as development proceeds. These are generally more detected by users of statistics than those who produce them. Regular dialogue between users and producers would reveal these gaps. A good place in detecting some of the gaps and deficiencies in agricultural statistics would be in the compilation of national food balance sheets.

For the collection of these statistics, it is not necessary to start from Stage I. It is a good working proposition if these statistics are put into the same stage as that prevailing for existing statistics, so that all statistics of the country could be uniformly developed.

CHAPTER 3

Agricultural Production - General Ideas

3.1 Importance

Let us apply the methodology described in Chapter 2 to current statistics in which annual agricultural production is the most important. Reliable estimates[1] of annual agricultural production are assuming a rapidly growing importance when countries are making serious efforts to tackle the feeding of their population, raising their nutritional levels and for planning national economies to raise their standard of living. In addition there is a growing demand to have the figures at lower levels too for planning and development of agriculture.

For convenience the items of production will be grouped into:
    (a) Crops – temporary and permanent;
    (b) Others – livestock and livestock products.

3.2 Crops

Crop production is dealt with here; livestock and livestock products will be dealt with in Chapter 7.

Production can be measured directly when the cultivated land consists of large estates or where the whole production passes through a few centres. Tea estates in India and Sri Lanka and rubber estates in West Malaysia are examples of the former. Cotton cultivation in the Sudan is an example of the latter where the entire cotton produced passes through 5 or 6 ginneries. In these cases accurate data of production are available. But

---

[1] Production, when made on, or after harvest, is known as an estimate and before harvest a forecast.

Agricultural Production — General Ideas

in the developing countries, agricultural holdings are numerous and small and it is therefore not practicable to measure or estimate production directly. We therefore estimate production indirectly by breaking up production into its two components area (cropped) and yield rate (per unit of land) and estimate each separately. This is the general method for temporary crops (such as rice, maize, wheat, soyabeans, peanuts, potatoes, cassava, vegetables). For permanent crops (such as coffee, coconut, rubber, fruit trees) instead of area, we take the number of bearing trees; and yield rate refers to per bearing tree. For permanent crops, area under crop is not meaningful as the number of trees per unit of land differs from holding to holding and from region to region. Area therefore is replaced by tree; actually it is the bearing tree that contributes to the production and thus area is replaced by bearing tree.

Development of production statistics would refer to the development of the two components, area and yield rate when temporary crops are considered. The three stages of development referred to in Chapter 2 when applied to production could be described as follows:

Stage I $\begin{cases} \text{Area determined subjectively} \\ \text{Yield rate determined subjectively} \end{cases}$

Stage II $\begin{cases} \text{One component determined subjectively} \\ \text{Other component determined objectively} \end{cases}$

Stage III $\begin{cases} \text{Area determined objectively} \\ \text{Yield rate determined objectively} \end{cases}$

In Stage II there are two possibilities. If we have our priorities right as enunciated in Chapter 2 then it should be area estimates are determined subjectively and yield rates determined objectively. The three stages will then be:

Stage I   Both components determined subjectively

Stage II $\begin{cases} \text{Area determined subjectively} \\ \text{Yield rate determined objectively} \end{cases}$

Stage III   Both components determined objectively

Yield rates and areas are dealt with in Chapters 4 and 5. Consideration is now given to direct estimation of production.

3.3 Direct Estimation

Direct estimation of production will come under any one of the three categories:

(i) Consisting of large estates with their processing plants meant for processing their products only;

(ii) As in (i) and in addition, processing the products of the small holders in the neighbourhood;

(iii) Consisting of only processing plants through one of which any production of the commodity concerned will have to pass through.

These concerns are like farms in the developed countries where those who control them are educated and knowledgeable persons, who can maintain good records. Those in categories (i) and (ii) in addition to accurate records of production maintain other records on agriculture.

These concerns or agencies are generally very cooperative to collect whatever data is requested by the Government. The Governments, however, have the necessary statutory authority to order the agencies to collect any data required for the nation.

As an example let us take a large tea estate in Sri Lanka and describe how production is recorded. The estate has its own factory where black tea is made from green leaf. The factory processes the green leaf of its own estate and that of the small holdings attached to it by statute for this specific purpose. Green leaf plucked on the estate is weighed twice daily and recorded – at noon and in the evenings. The weighing is done in the presence of the plucker and there is no possibility for any faulty recording. In the case of the small holdings, green leaf is taken over by the estate once daily and the weighing is done in the presence of the producer. There is thus a daily record of production of green leaf of (1) the estate and (2) the small holdings – recorded to the nearest pound. As regards black tea produced, there is a running record in the factory of made tea every 24 hours – recorded to the nearest pound. The production data thus collected are sent periodically to the Government Tea Commissioner who has statutory powers empowering him to collect such data. These data would find their way into the various Government Statistical Bulletins through the Census and Statistics Department which is in overall charge of all Government statistics. The system prevailing in these estates can be

Agricultural Production – General Ideas

considered satisfactory. In a case like this, there is no need to do anything further for data collection except to periodically examine the data collection and modify the questionnaires sent to estates to meet the changing needs.

Let us take another example – with sugar plantations in Mauritius. As in Sri Lanka for tea, in Mauritius there are large estates with factories and small holdings without factories. In the case of tea, production is recorded almost daily throughout the year, whilst for sugar cane it is limited to some months of the year when harvesting of the cane takes place. In Mauritius, estates maintain good records of sugar cane (i) harvested in the estates and (ii) bought from the small holders. Further regular records are maintained for raw centrifugal sugar produced by the factories. The authority which collects the data from the estate is the Sugar Insurance Fund Board. As this Board pays compensation when the sugar cane is damaged (mainly by cyclones) all producers, big or small, register their names with the Board, ensuring almost 100% coverage. The prevailing system for collecting data on production of this crop can be considered satisfactory and further needs can always be met by the Board with the assistance of the estates.

The two examples given are both from small and compact countries. If, however, a country is large like India or scattered like Indonesia, there is a time lag for the competent authority to receive the returns from the estates. This can be overcome by direct contact between the competent authority and the estates without any intermediary like a local authority coming in between them.

Production estimated directly as described here may or may not cover the entire crop concerned. Coverage is complete in the two examples given, namely tea in Sri Lanka and sugar cane in Mauritius because the finished product has to come out of the factories which are well known. This may not be the case for rubber where the conversion from latex to dry rubber is done by a large number of small curing plants. In such cases production is compiled for large estates directly and for small holdings by estimating (i) the two components area and yield separately and taking the product (this would be dealt with in later chapters) or (ii) production directly by sampling methods.

For small holdings method (i) is preferable when area and production estimates are needed. Method (ii) can be employed when only production estimates are needed. The frame for the sample would be the list of small factories or plants which should be periodically brought up to date. It is usual in such cases to have a stratified sample — stratified geographically for administrative needs and by size of the factory.

CHAPTER 4

Agricultural Production - Crop Yields

4.1 Introduction

Determination of crop yields is primarily for the estimation of crop production. There are other uses as well. At a glance they show the differences in the yield rates between different administrative units, ecological/agricultural zones and other units as provided in the stratification and design of the sample. They also show trends in improvement or otherwise over the years of the various cultural practices and other measures adopted for improvement of agriculture. Such information is useful in evaluation of projects and for taking further steps in the promotion of agriculture.

In our theme of development and statistics, we have mentioned the priority that should be given to the determination of yield rates in the estimation of production. This would take the form of using objective methods with sampling and physical measurement of the crop. The sample consists of several units which are termed "crop cutting plots". These are the plots which are harvested and the produce weighed. They are of prescribed dimensions located and marked according to clearly defined procedures. From the yield rates of these plots, yield rates could be determined with their standard errors for the larger units like villages, sub-districts, districts/regions and finally for the country. It is with this background that yield surveys (or crop cutting surveys) are undertaken in a country.

4.2 Crop Yield Surveys

An efficient design for these surveys has to take into consideration the administrative set-up of the country, availability of statistical staff, the nature and amount of information available with the help of which the sample is to be selected and funds at the disposal of the country for the survey. In other words we have to consider having a suitable frame for drawing the sample and design the most efficient sample within the framework of the country's existing services and funds. It will be convenient to deal with the subject under the following items:

(i) Preparation of a frame;
(ii) Stratification;
(iii) Method of sampling within strata;
(iv) Crop cutting.

4.2.1 Frame

For drawing a sample of plots for crop cut, we should have a list of all lands under the crop for the season. For lands cadastrally surveyed or with high density of cultivation, lists of cropped lands are normally available in developing countries. This is not so for non-cadastral lands and lands with low density population particularly in Africa where it is only possible to have lists of farmers through whom we locate the cropped lands. Such a list whether of lands or farmers is called a frame for the purpose of drawing a sample.

In determining yields, eventually we have to reach the crop cutting plots — either through a frame consisting of a list of fields cultivated with the crop or through a list of farmers cultivating the crops. Some countries have taken one or the other method for building the frame while others have had a combination of the two — applying one method for the organized sector (of lists of fields) and the other (list of farmers) for the non-organized sector.

Both methods have advantages and disadvantages. However, there is not much choice at the beginning because one has to use what is available before a proper frame is built. One has to examine whether earlier population or housing census results could be used to build a frame by extracting agricultural households. Tax lists could also be examined as well as lists of lands under irrigation. There are several sources available with Governments and Local Government bodies which should be scrutinized to make a start for building a frame. It is hardly necessary to state that a good frame is pre-requisite for a proper survey.

4.2.2 *Stratification*

Stratification is used in all large scale surveys. It has two purposes:
(i) To increase the accuracy of the overall population estimates;
(ii) To ensure that sub-divisions of the population which are themselves of interest are adequately represented (called domains of study).

Maximum overall accuracy can be attained if the strata are so chosen that the units within each stratum are as similar as possible and that mean values of the various strata differ as widely as possible. It is advisable to use domains of study as principal strata even if some other form of stratification might give more accurate results.

In yield surveys one has to provide not only an accurate and unbiased estimate of the yield rate for the country as a whole, but also similar estimates for the main administrative divisions. Generally these divisions which are of the third level of administrative hierarchy in a country are taken as domains of study (or principal strata).

Further stratification within the domains of study is advantageous in increasing the precision of the yield estimates. The rate of yield is known to differ widely from place to place and therefore a geographic stratification suggests itself. This can be done by either sub-dividing each domain into compact arbitrary areas or into existing administrative sub-divisions. The latter course is preferable in drawing the sample and organizing the field work. Another possible stratification is by ecological or agricultural zones or by traditional and improved cultivation. It has been found convenient for paddy cultivation is mountainous regions to sub-divide the domain of study into hilly areas and valleys.

The method of allocation of sample units between different strata has to be considered next. The most efficient distribution of the total sample among the different strata would be the one for which an estimate with desired precision is obtained at minimum cost. Where locally stationed field staff are available for crop cuts, the cost would mainly depend on the number of crop cuts to be sampled. Under this condition, the question of optimum allocation between strata reduces to that of attaining maximum precision with a given number of crop cuts. Neyman has shown that the total number should be distributed among the different strata in proportion to the product of crop area and standard deviation of the yield of the respective strata. It has been found from experience that values of standard deviation do not differ significantly for different strata within an administrative district which we have taken as the domain of study. Hence the distribution of crop cuts in proportion to the area under the crop in the strata, could be considered as a close approximation to the optimum allocation.

Though every effort should be made to distribute the crop cuts in proportion to the area under the crop in the different strata, it is not always possible to strictly adhere to this distribution. It should, however, be borne in mind that any serious deviation from the principle of proportional allocation reduces the statistical gain from stratification appreciably. For example, the overall precision of the district (domain) estimate would be unfavourably affected when deviating from this principle a certain minimum number of crop cuts is sampled within each stratum for which a separate estimate of yield is desired with a given level of accuracy. So would be the case when there is an upper limit to the number of plots that can be sampled in any stratum, irrespective of its crop area, due to practical problems in the field. Such cases of deviation from proportional allocation are inevitable.

When stratification is done strictly in proportion to the crop areas, the estimation of the average yield rate for the domain as well as its precision is simple as these become exactly analogous to the corresponding estimates from unstratified simple random sampling. Where stratification is not done in proportion to the crop areas of the different strata, the estimates of yield rate and their precision have to be calculated separately for each stratum and then combined into corresponding estimates for domain by weighting them with crop areas in the different strata.

4.2.3 *Method of sampling within strata*

It is not practicable to select the crop cutting plots in a single process; it has to be devised in successive stages. Organizational convenience and efficiency of field work require that the sampling units be natural units or groups of units (particularly for regular surveys) rather than arbitrary units selected from maps by drawing beautiful parallel, vertical and horizontal lines.

Normally the crop cutting plots are selected by a process of a 3 or 4 stage sampling (called multi-stage sampling). Where villages are well-defined these are suitable as primary or first stage sampling units. A farm may be taken as the first stage unit if the country's agricultural land consists of large farms. The second stage units are either fields where the frame consists of areas of land, or farmers where the frame consists of farmers cultivating the crops. The third stage units would be the crop cutting plots. What has been stated applies to temporary crops. For permanent crops the second stage units are stratified into bearing and non-bearing and the third stage units which are clusters of trees (and not plots) are selected from the bearing trees.

It should be mentioned that with a given sample size we obtain maximum precision for a single stage selection of sample. As there are practical difficulties for a single stage selection resort is made to multi-stage. However, the number of stages should be kept as low as possible.

In selecting the sample units at the several stages, use is made of random numbers. This is done by numbering the units of the population concerned and drawing the requisite number for the sample from a table of random numbers. This renders the selection a random one as the method gives an equal chance of selection to all numbers of the population. In the actual process of selection, the first stage units are drawn with the help of random numbers from a complete list of units available with the administration, and the second stage units are drawn from the selected first stage units after preparing lists of fields growing the crop or the lists of farmers as the case may be. (These selections are done by the statistician in charge of the survey.) The third stage units are drawn by the field workers from the fields selected (these fields are selected directly where the frame consists of areas of land; where the frame consists of farmers the fields will be those of the selected farmers; for permanent crops fields selected are from those with bearing trees). Plots (for temporary crops) and clusters (for permanent crops) are located in the selected fields with the help of random coordinates taken from two adjacent sides of the selected fields.

In the method described we have followed simple random selection of villages/farms and fields. This method gives smaller villages/farms and smaller fields a greater chance per unit area under the crop of being included in the sample than larger villages/farms and larger fields. If the yield rate is correlated with the size of village/field, this method of selection would lead to a biased estimate, when a simple arithmetic mean is calculated from the sample. The Indian experience has shown that there is no correlation between size of villages and of fields on one hand and yield rate on the other and this method has been suggested on this experience. Perhaps many countries share this experience as regards correlation between size of fields and yield rate. If, however, any country finds a contrary experience then the sample units will have to be selected with probability proportional to size.

Having distributed the sampling units among the different strata according to the area under the crop as described earlier we shall discuss what is the best practical distribution of the sample among different stages. Considerations of precision suggest distribution to every primary unit (village/farm). However, there are other considerations (like movements of field staff, poor communications, limited harvesting time) which would force one to reduce the number of primary units and correspondingly increase the second and third stage units. As to the number of villages to be sampled it can be worked backwards by determining the number of second stage and third stage units. Experience in many countries has shown that variation in yield rate between fields is far greater than within the same field. So it is wise to select more fields and restrict the number of plots per field in the sample. It is generally the practice to select two plots per selected field at the first survey (to provide a basis for evaluating the variability within fields) and thereafter restrict to one plot per selected field and two or three fields per selected village. Indian experience shows that two to three fields per village with one plot per field is about the optimum distribution.

4.2.4 Crop cutting

It is convenient at this point to state briefly the various steps taken to get to the crop cutting plot within the domain. The domain, which is

Agricultural Production – Crop Yields

generally a principal administrative district is stratified into several sub-divisions called the strata. The number of plots assigned to the domain is distributed into the various strata in proportion to the areas under the crop in the different strata. If three fields are taken in the second stage and one plot in the third stage, the number of villages for a stratum would be one third of the number of plots allotted to the stratum. This number of villages is selected either at random[1], or with probability proportional to the areas under crop in the villages in the stratum as stated in section 4.2.3. In each selected village a complete list of fields under the crop is prepared where the frame consists of lists of areas. From this list three fields are selected at random and from each field one plot is selected at random.

If the frame consists of lists of farmers, then the farmers can be contacted for the selected villages and lists of their fields prepared and three fields are selected at random as before. If, however, this is difficult, three farmers are selected at random for each selected village and lists of their fields prepared and three fields are taken at random – in this case the sampling becomes a four-stage one.

For permanent crops, at the second stage where the frame consists of lists of areas these areas are stratified into bearing and non-bearing trees and three fields are selected from the bearing fields where the frame consists of lists of farmers. Thereafter, the procedure stated earlier in this sub-section is suitably modified.

Size and shape

During the last three decades attempts were made with circular, triangular, rectangular and square plots. Now most countries have settled down to rectangular or square plots.

Plot sizes vary according to the crops and country. Asian countries generally have plot sizes varying between 1/80 and 1/100 acre (40 and 55 sq. metres). In other countries even half this size is used. Generally there is an over-estimate in very small plots and this decreases as the plot size

---

[1] Where there is no correlation between size of village and yield rate.

increases, e.g. in India 1/100 acre plots gave an over-estimate of 0.8% only for paddy. It is therefore essential that plots should not be very small.

Harvesting and weighing

Precise instructions have to be given to the field staff for marking and harvesting plots as biases arise due to deviations from the instructions given. Two further factors have to be borne in mind. Firstly the field officer must harvest (and thresh where necessary) in the same manner as the farmer does otherwise he would be measuring the *biological yield*[1] and not its *harvested yield* which is the yield the farmer measures. Secondly in some cases the sample includes either harvesting immature grain or excluding from the sample later maturing fields (where the field staff is a moving one). This can be circumvented by having locally stationed field staff for crop cut.

If the weighing of the produce is done by the farmers at the time of harvest then the field officer has to weigh the produce from the crop cut in the same way. If the weighing is done after some days, then some arrangements have to be made by the field officer to preserve the produce (after labelling) and weigh later. This is not a very practical approach. In such cases it is better to weigh the produce at harvest time and correct the weight by conducting some driage trials on a sub-sample basis with a standard moisture content.

An adjustment to the yield estimate is necessary in all surveys on grain crops to allow for the moisture content of the grain. The concept of yield rate is that when multiplied by area harvested would give the harvested production at farm-gate.

4.2.5 Estimation of mean yield

It is assumed here that the domain is an administrative district and stratified by sub-administrative divisions. Within such a division (stratum) a number of primary or first stage sampling units (villages) would be

---

[1] Biological yield is the theoretical yield — without allowing for harvesting and threshing losses.

Agricultural Production – Crop Yields

selected approximately proportional to the area under the crop. From a selected primary unit a fixed number of second-stage units (fields under the crop) would be selected and finally from each selected field, one crop cutting plot of prescribed size and shape would be selected as described earlier in the chapter.

If there is evidence of correlation between yield rate and size of sample units, the first-stage and second-stage units are selected in proportion to their crop areas. The crop cutting plot, however, is selected at random. Then the estimate of the mean yield rate for the stratum $\bar{y}_s$ is given by

$$\bar{y}_s = \frac{\sum_{i=1}^{n} \sum_{j=1}^{m} y_{ij}}{nm}$$

where
- y_{ij} denotes the plot yield from the jth field of ith village,
- m denotes the number of fields selected per village,
- n denotes the number of villages in the sample.

This estimate is unbiased.

This simple arithmetic mean will be a biased one when villages and fields are selected with equal probability by simple random sampling. This estimate of the mean, however, has been found an acceptable one in India and some other countries in view of the absence of correlation between yield rate and size of villages/fields.

If the simple arithmetic mean is biased for selection with equal probability then an unbiased estimate is given by

$$\bar{y}'_s = \frac{1}{A} \frac{N}{n} \sum_{i=1}^{n} \frac{M_i}{m} \sum_{j=1}^{m} A_{ij} y_{ij}$$

where
- A is the total area under the crop in the stratum,
- A_{ij} is the area of the jth field in the ith village,
- N is the total number of villages under the crop in the stratum,
- M_i is the number of fields under the crop in the ith village.

28 Agricultural Statistics: A Handbook for Developing Countries

Experience has shown that when a simple arithmetic mean does not provide a good estimate on account of its bias, the ratio estimate given by

$$\bar{y}_s'' = \frac{\sum_{i=1}^{n} A_i \frac{\sum_{j=1}^{m} A_{ij} y_{ij}}{\sum_{j=1}^{m} A_{ij}}}{\sum_{i=1}^{n} A_i}$$

though not strictly unbiased, is a better estimate than the weighted estimate given earlier.

Once the strata means are estimated, the estimate of the domain is obtained by weighting them with the respective crop areas.

4.3 Example (Temporary Crop)

An illustration is given here of a yield survey; maize in Egypt. Egypt is divided into three large geographical regions known as Upper Egypt, Middle Egypt and Lower Egypt. Each of these regions is divided into several provinces and provinces into districts and sub-districts. Generally provinces are taken as the domains of study. When a crop is not an important one then the domain of study is one of the three main regions of the country. For maize provinces are the domains of study in Middle and Lower Egypt where the area under maize is large and consequently a separate estimate of yield rate of sufficient precision is required for each province. The area under maize in Upper Egypt being small, this region is treated as a domain of study and a single estimate is sufficient for Upper Egypt. In all twelve separate estimates (yield rates) are needed, one for Upper Egypt, four for Middle Egypt and seven for Lower Egypt.

In the survey, each province was stratified into districts and each district into two or three agricultural units so that these agricultural units served as strata for the province. In each stratum the cultivated land falling in each village was divided into clusters of about 200 feddans (range 150–250 feddans) — these constituted the *primary* sampling units. From a

Agricultural Production – Crop Yields

	Area under maize 100 feddans*	Number of clusters selected	Number of crop cutting plots
UPPER EGYPT	628	120	240
El Minya	1394	100	200
El Fayum	1089	100	200
Beni-Suef	1172	100	200
El Griza	668	100	200
MIDDLE EGYPT	4323	400	800
El Qalyubiya	998	100	200
El Minufiya	1979	200	400
El Sharqiya	3062	250	500
Daqahliya	1666	130	260
Kahr El Sheikh	1028	100	200
El Grharbiya	2160	200	400
El Beheira	1899	200	400
LOWER EGYPT	12792	1180	2360
EGYPT	17743	1700	3400

Source: Koshal, A review of work done in U.A.R., 1960.
*Feddan = 4200 square metres.

selected cluster, two parcels[1] growing the crops were randomly selected and from each parcel one field was selected at random and from the selected field one plot of size (7m x 6m = 1/100 feddan) was selected at random.

Of the 3400 plots, 240 were allotted to Upper Egypt and balance to the provinces in Middle and Lower Egypt as in the table. The assigned number of clusters in each province was distributed to the various strata in proportion to the area under maize in the strata.

A crop cutting plot was located in the following manner. Assume the selected field as rectangular in shape. Locate the south west corner (SWC)

---

[1] Parcels consist of fields and fields consist of plots.

of the field (any corner would do) — from this corner if you face north the field is on your right-hand side. Measure the length, l and breadth, b of the field to the nearest metre. Select a random number (l_1) between 0 and l-7 and another number (l_2) at random between 0 and b-6 (7 and 6 being the length and breadth of the plot). Then l_1, l_2 are the coordinates of SWC of the plot. From this point on the ground with 4 poles planted at the four corners of the plot, run a string round the plot. (A check on the correct marking of the plot on the ground is to measure the diagonals of the plot which should be 9.22 metres.)

The harvesting was done in exactly the same way the cultivators harvest in the locality. The plants were cut by boys and girls with the scythe and cobs separated from the plants. The cobs were exposed to the sun for half an hour on a hessian cloth and thereafter counted and recorded.

For driage tests, sub-samples were taken and allowed to dry for a fortnight. Thereafter the grain was taken from the cobs and weighed to the nearest 10 grams. From these recordings the yield rates at various levels[1] were calculated.

The following observations may be made on the survey:

(i) The survey illustrates that the domains of study need not be uniform for the whole country; provinces are suitable where the crop is extensively grown and an entire geographic region (comprising of several provinces) where the crop is a minor one;

(ii) Though villages are quite suitable as primary sampling units in many countries, the survey illustrates how successfully clusters of cultivated land of about the same size can be formed for the same purpose. This, however, is possible only where cultivation is done in large tracts and cadastral maps of the area are available as in the case of Egypt;

(iii) The plot size has been found suitable; not too small for large bias in the yield and convenient for calculations (one has merely to add two zeros to obtain per feddan yield).

The methodology described here for maize could equally apply to other temporary crops with suitable modifications.

---

[1] Levels of reporting — see section 5.2.2.

4.4 Example (Permanent Crop)

Another illustration is given for yield surveys; in this case coconut in Sri Lanka (Ceylon). The survey was carried out in 1962/3 for determining yields and area under cultivation with the assistance of Mr. K. V. R. Sastry, FAO statistician. A modified version of the survey is given here excluding estates. The frame consisted of:

(i) Small-holdings: holdings[1] less than 20 acres but more than one acre;

(ii) Gardens: holdings of one acre or less.

The survey was restricted to 3 of the 22 administrative districts of the country. These 3 districts are contiguous and covered two-thirds of the total area under the crop. The districts were the domains of study. Each district consisting of several DRO divisions[2] were used as geographical strata. As villages were found to be extremely heterogenous in terms of total cultivated area, survey blocks were formed having about 500 acres per block. 1389 blocks were formed from 5073 villages and these blocks were used as *primary* sampling units.

120 blocks for the survey were allotted to the three districts (45 in Colombo, 55 in Kurunegala and 20 in Chilaw). Allocation of sample blocks among strata in each district was roughly proportional to the corresponding coconut area (excluding estates) from the 1946 Census of Agriculture. In each sample block a complete list of all agricultural plots was prepared according to the physical lie of the land. Any plot consisting of at least eight coconut palms was considered as a coconut plot and classified into a small holding plot or a garden plot.

In each sample block 10% of small holding plots and 4% of garden plots were selected for the *second* stage of the survey which consisted of physical enumeration and classification of palms according to 6 age-groups and 5 bearing-status classes. Small-holding and garden plots were selected systematically in clusters with a random start; each cluster consisting of 5 and 2 consequently listed plots for small-holding and garden plots respectively – we shall call the selected plots sample plots.

From these sample plots, 3 small-holding plots in each sample block

---

[1] Holdings: see chapter on census of agriculture.

[2] These are sub-districts.

and a sub-sample of 2 garden plots in alternate sample blocks were selected at random for *third* stage of survey pertaining to collection of objective data on yield — we shall call the selected plots sub-sample plots.

There is a fourth stage in the survey involving a further selection of eight bearing palms in two random clusters[1], each with 4 adjacent bearing palms,[2] within each sub-sample plot to record data on actual number of nuts plucked at successive harvests for the period of survey (July 1962-June 1963). It is from these yield records, yield rates were calculated for the survey.

For conversion from the number of nuts to weight in unhusked nuts, and husked nuts, nuts plucked from each of the two key palms that defined the clusters were husked at the time of the visit of the field officer and weight recorded to the nearest 1/10 pound of unhusked nut and husked nut.

The methodology described here for coconuts could equally apply to other permanent crops with suitable modifications.

---

[1] These clusters correspond to the crop cutting plots for temporary crops.

[2] For each cluster of 4 palms, 1 palm (called the key palm) is selected at random from the sub-sample plot and 3 adjacent palms are taken to form 4.

CHAPTER 5

Agricultural Production (Crops) - Area

5.1 Introduction

One component of production, yield rate, has been dealt with in Chapter 4. This chapter is concerned with the other component — cropped area for temporary crops, with modifications for permanent crops.

Most developing countries collect data on cropped areas on a system of reporting, starting from the lowest levels of administration and ending up at the national level. The system, known as the reporting system, generally depicts the level of development of area statistics in the country. It is therefore useful to describe the reporting system as prevailing in the countries as a starting point for the subject.

5.2 Reporting System

In developing countries land tax has been in vogue for a very long time. In administering tax collection, returns are sent by tax officers at the village level to their superior officers and these officers in turn to their superior officers till summaries are made available at regional/provincial/national levels. When the time came for collecting crop areas on a systematic basis after World War II the Governments made use of these returns to extract the needed data. In due course special forms were prepared for this purpose and used by officers at the village level, to collect the data needed on areas cropped and other data on agricultural statistics.

Forms used and methods adopted in the reporting system vary from country to country and in large countries from region to region. The reporting starts from a village officer or a technical officer (like an agricultural or statistics officer) who is in contact with peasants. The data

34   Agricultural Statistics: A Handbook for Developing Countries

would consist of monthly or seasonal, planted and/or harvested areas, of main crops and such other information requested by the national authority on agricultural statistics. The data collected would pass through the various levels of administration[1] (3 or 4) to get to the national level where they are processed and made available to the Government and the public.

If we examine the data as collected we will generally find that the quality depends on many factors such as the level of the reporting unit, the method of collection, the availability of cadastral maps and whether the estimates are obtained by subjective or objective methods.

The most important aspect in the system is the level of the reporting unit. The unit may vary from farm/field, village, district to province. It is reasonable to expect that the most accurate data will be obtained when the farm/field is the primary unit. Countries with farm/field as reporting units will have more dependable statistics than those with larger units. It can hardly be expected that anybody could provide accurate estimates about a large administrative unit at one time. It is only those countries with lower reporting units that have improved the accuracy of their estimates by using random sampling techniques with physical measurement. With large units, it is difficult to use these techniques.

Next in importance is the method of collection. If the reporting unit is a farm or field, the Government officer in charge of collection of data either gets it from the farmers or makes spot inspections. If the farmers are knowledgeable the data would be dependable, otherwise spot inspections would be necessary. Experienced officers can check on data collected by inquiring on the amount of seed used as specified areas and geographic units use standard amounts of seed per unit of land (hectare, acre). When the reporting unit is not a farm or field but a bigger unit, reporting becomes difficult. The officer concerned makes an estimate of the area (a) by inquiring from a few farmers (not necessarily representative) or (b) by visits to the area. The first method is not satisfactory as the farmers are not randomly selected and even if randomly selected, the estimate would suffer from the inability of the farmers to give accurate figures. The second method would be accurate only if the cropped area is in one block (which is not often the case), and some map or plan is available indicating

---

[1] Sometimes referred in the text as "levels of reporting".

the rough acreage. The countries using big reporting units are fast realizing their unsuitability and are gradually replacing them by smaller units.

Where lands have cadastral maps, reporting is obviously more dependable than where there are no cadastral maps. If for cadastral lands, the reporting unit is the farm or field and spot inspections are made by an experienced reporting officer, the accuracy of the data collected is greatly enhanced. Further improvement could be made if measurements are made on farms/fields on a sampling basis and estimates modified with the use of suitable correction factors evolved from these measurements.

Development of cropped area statistics would now be treated on the lines suggested in the preceding sub-paragraph. It is convenient to treat cadastral and non-cadastral areas separately after some general remarks common to both.

5.3 Guidelines for Development

5.3.1 *General*

Most developing countries collect current area statistics by the reporting system. In the process of improvement of these statistics by the use of objective methods, it is not necessary to do away with the system. On the contrary the system can be used with advantage introducing some objective methods into the system to improve on the accuracy of the estimates.

In the collection of cropped areas, greater frequency and more details are needed for major crop than for minor crops. For paddy which is a major crop in Indonesia, monthly data on standing crops at the beginning of the month, crop harvested during the month, new plantings for the month are collected, while for minor crops and vegetables, data are collected only quarterly.

Concepts and definitions should be without ambiguity for the terms used. Data on areas generally refer to gross areas, i.e. inclusive of bunds, ridges and channels within the field. For estimating production we need net areas — these are generally obtained as a percentage of gross area from sample surveys by physical measurements of the areas concerned. In mountainous regions, areas refer to projected areas — from a practical point of view, however, this is not a serious issue as even a 25% slope gives a 5% difference between projected area and actual area. It is useful to fix a minimum area for each crop for which data would be collected.

5.3.2 Cadastral areas

The first priority in development is to have the farm or field as the lowest reporting unit. Countries which do not have farm or field as the lowest reporting unit should without further delay remedy this. There is no alternative to this, to provide accurate area estimates. With cadastral lands, the cost involved in switching over to farm or field as reporting unit, would be within the means of most countries. Further, when the time comes to use objective methods with measurements of areas, it is advantageous to have small reporting units.

There is no guarantee that the adoption of small reporting units alone will necessarily mean an improvement. Other measures should be taken along with small reporting units. This brings us to the next step which is the method of collection. With farm/field as the unit, data can be had from the farmer or by field inspection. With cadastral maps and knowledge of local agricultural conditions, most officers at village level can collect dependable data from the farmers. The ideal would be a farm-to-farm inspection or a field-to-field inspection. This, however, is not a practicable proposition in all cases with many crops and frequent reporting. In such cases one detailed visit followed by periodical visits will help to check on the data supplied. When whole fields are under one crop, with a cadastral map it is quite simple to record the area. But where part of a field is under one crop and the balance under others, care should be taken in estimating the cropped areas separately.

With farm/field as the reporting unit and data are collected on this basis, it is not possible to do the work without an officer being available at the village level. This collection of data has been part of the duties given to village headmen in the past and recently going into the hands of cooperative societies in some countries while other countries are retaining the headmen system. The present system has to be continued as it is not possible for developing countries to find statistically trained staff at village level. (Countries have statistically trained staff only at sub-district level.)

It is useful for the village officer to maintain a register for agricultural statistics. Areas cropped and other data could be entered periodically and kept up-to-date. The author has seen some fine records maintained (in registers) by headmen in Java (Indonesia). From this register either the headman sends the required data to the agricultural/statistical officer of the sub-district level or one of these officers calls at the headman's office to collect the data. Maps should be maintained at these offices and portable maps taken to the field during inspections. As the lands are cadastrally surveyed these maps would be of immense assistance to assess accurately the cropped area. Coverage should be made as complete as possible by enumerating all the cultivated lands or all the farmers in the villages depending on the system in vogue.

The data on cropped areas at the lowest level are assembled at the office of either the village officer or sub-district officer. The data collected are sent to the district office after scrutiny and some test checks in the field by the sub-district officer. At the district office the data for the several sub-districts are assembled and the district data are sent to the national office (through a provincial office if there is one, otherwise direct). In this system there is some delay for the national office to receive the data. If statistics are to play a role in its usefulness, timeliness is essential. To avoid delays, some countries have taken a wise step in requesting for copies of reports from sub-districts direct to the national office while awaiting the official data through the normal administrative channels. Administrative steps such as this, are as essential as technical improvements to the development of statistics.

In the reporting system, stress is generally laid on data collection in the field. Equally important is the part played by the national office. For preparing the several questionnaires, maintaining uniformity in collection of data, training and supervision of field work and scrutiny and analysis of data, the national office has a heavy responsibility. Questionnaires have to be periodically reviewed to meet the ever increasing demands of users and for further progress. Proper instructions and periodical training should be given to all those involved in the collection and assembling of data including field training. Such steps should ensure uniformity in reporting. Supervision at every stage of collection of data is an essential inclusive of test checks in the field.

In improving the quality of any statistical data, objective methods are generally advocated. For determining yield rates objective methods have been insisted due to large errors involved in the use of the alternative subjective methods. In estimating areas in cadastral lands by the reporting system the error is not likely to be serious if the improvements suggested in this section are carried out.[1] In view of this and in view of the high cost involved for entirely replacing the reporting system by sample surveys using objective methods, most countries have found it most advantageous to use a combination of subjective and objective methods. In other words we basically use the reporting system and improve the estimates by sample surveys using physical measurements. In the data collected (gross areas) by the reporting system two corrections are necessary:

    (a) Due to defects in the gross area caused by factors such as minor crops and large bunds and drains being included in the reported area;

    (b) Due to field drains, channels, bunds reported in the gross area but not included in the plots for crop-cutting.

It is here that objective methods are suggested on a sampling basis in the clusters/fields selected for crop cutting. When crop cuts are conducted for determining yield rates, those clusters/fields from which crop cutting plots are selected are measured as described in section 5.4. The cropped area in the entire cluster/field is measured as well as the area under drains, channels, bunds within the cropped area. These should be calculated and correction factors worked out for each domain of study. As an example, let us take the sample survey for cotton in Egypt 1956. For the province Benisuef the estimate from the reporting system and the sample estimate of gross area were about the same: while for province Sohag, the estimate from the reporting system was an under-estimate by 4.5%; while the correction factor due to field drains, and channels to reduce to net area was estimated at 1.4%; consequently the overall correction factor was 3.1%.

Such sampling methods for improving area estimates are done along with crop cut for economy of time and money. These methods are

---

[1] In India, experience has shown that official estimates of crop acreages obtained through complete enumeration (reporting system) in cadastral areas by the headmen, are reliable and can be used for all policy decisions, even for smaller administrative units like districts or sub-districts.

preferable to total replacement by objective methods as experience has shown in India and other countries that with the usual sample size for such surveys estimates for areas were not found satisfactory even at national level. To get 0.5% error for the whole country and for provinces between 1% and 2% Egypt had to take a 25% sample which undoubtedly is costly. All these go to show that the use of entirely objective methods using sampling to provide estimates for small administrative units is beyond the resources of most developing countries and not necessary.

Summing up for cadastral areas, development lies in the improvements to the reporting system and periodical sample surveys using objective methods to improve on the quality of data collected from the reporting system as detailed in this sub-section.

5.3.3 Non-cadastral areas

Non-cadastral areas present difficulties in the collection of data of cropped areas due to non-availability of accurate land extents at the farm/field level. The long term answer lies in having the lands cadastrally surveyed when all the steps suggested in section 5.3.2 could be taken for improvement of the quality of data collected. In the long term development plans of the Governments, provision should be made for cadastral survey of these lands over a period of time. In the meantime statistical staff with some training on area measurements could survey the cultivable lands of the country on a systematic basis spread over several years with simple equipments details of which are given in section 5.4. Maps prepared by such surveys, though not as accurate as the ones provided by qualified land surveyors, would serve our purpose for reporting on cropped areas on these lands. Gradually when parts of non-cadastral lands are surveyed and cadastral maps are made available, steps for improvement for such lands should be as in section 5.3.2.

Until such time non-cadastral lands are cadastrally surveyed most of the steps suggested for improvement of the reporting system could be taken, which, however, would not be as profitable as with cadastral lands.

5.4 Area Measurements

5.4.1 General

For very accurate area measurements, trained surveyors and expensive equipment are needed. Not many developing countries can afford such high expenditure for statistical purposes and even when they can afford it takes considerable time to cover the whole country. We have therefore to resort to cheaper and quicker methods which the statistical field staff can apply without much difficulty.

Area measurement is quite simple when maps and plans of the ground to be studied are available. Some Governments keep such maps up-to-date at least for some parts of the country by the cadastral services, generally for purposes of revenue. Cadastral maps show the boundaries and area of each parcel of land together with the name of the owner.

The situation we meet in many developing countries is that cadastral maps are not available for the whole agricultural area or that the maps are not up-to-date. The Agricultural Statistician has therefore to obtain in the field the data he needs.

In our work we resort to area measurements in the following cases:

   (i) Cropped areas;
   (ii) Non-cadastral lands;
   (iii) Holdings.

In cropped areas, measurements are done on a sampling basis (a) to improve on the quality of data collected by the reporting system and (b) to estimate cropped areas objectively as an alternative to complete enumeration by the reporting system.

In non-cadastral lands, measurements are done on a systematic basis to provide in due course cadastral maps of cultivable lands in such areas for the whole country as an alternative or forerunner of such maps by trained surveyors.

Holdings are measured on a sampling basis in agricultural censuses as a check on the data provided by the holders.

There may be other cases beyond these three mentioned for measurement of areas.

5.4.2 Procedure

When a parcel or a field is to be measured, it is necessary to get a rough figure of the parcel/field concerned on a convenient scale and then proceed to measure. Most parcels/fields have geometrical figures in the forms of polygons; if not these have to be drawn to the nearest polygon. The measurements then consist of measuring the sides and angles of the polygons concerned. If the figures have few sides like quadrilaterals it may be advantageous to break them into triangles and measure the sides of the triangles, as errors are less with sides than with angles in measurements. It is, however, not possible to avoid measurements of angles in some cases on level lands and slopes in hilly areas.

In measuring distances,[1] the following methods/instruments are used:

(a) Pacing;
(b) Pedometers;
(c) Topofil;
(d) Tapes, chains;
(e) Optical distance measures/rangematic instruments.

Which method/instrument is to be adopted/used has to be settled by the country depending on the availability of funds, level and skill of the field staff and terrain of the land.

In measuring angles, many types of compasses are available in the market[2] — for greater accuracy, magnifying prisms are used. Such compasses are called prismatic compasses.

After the field work is over, the parcels/fields should be drawn to scale in office. If the field work is done properly most of the diagrams will close. But cases will arise that the diagrams do not fit together. Corrections for such diagrams are described on page 55 of Zarkovich's book on *Estimation of Areas in Agricultural Statistics*.

Calculation of areas can be made by triangulation using the well-known formula for the area of a triangle $\sqrt{s(s-a)(s-b)(s-c)}$ as any polygon can be

---

[1] For details, see Zarkovich's *Estimation of Areas in Agricultural Statistics*, Rome, 1965, pages 38-40.

[2] *Ibid*, pages 40-45.

made up of triangles. For easy calculation of areas, an instrument called a planimeter is used.[1]

5.5 Priorities

In estimating cropped areas, priority is given to non-cadastral areas over cadastral areas, as errors in estimation from the former are greater than from the latter. In non-cadastral areas, proper surveys for providing cadastral maps for cultivable lands take first place. This has to be spread over several years as the work is heavy and funds needed are large.

The next priority is to conduct sample surveys on cropped areas on cadastral lands to improve on the estimates obtained from the reporting system.

5.6 Mixed Cropping

Mixed cropping is found in many developing countries where two or more crops are grown simultaneously in such an intermingled fashion that it is difficult to identify any set of pattern of cultivation. The composition of a mixed crop field may vary from time to time since not all the crops in the mixture may be planted together or harvested together. The constituent crops may have unequal growing periods and different harvesting frequencies. For example, in a mixture of maize and cassava, the maize would be harvested in 4 or 5 months completely while cassava would take 9 to 20 months.

Mixed cropping can involve extremely methodical systems of intercropping with the different crops arranged in orderly fashion (alternate rows or groups or plants). In the alternative it may be a completely haphazard mixture of plants. In some cases the different crops have equal status and in other cases one crop is the principal or primary crop while the others are secondary. The presence of one crop in some cases may have little or no influence on the others while in other cases the yields of the constituent crops affect each other.

[1] For details, see Zarkovich's *Estimation of Areas in Agricultural Statistics*, Rome, 1965, pages 58-60.

Agricultural Production (Crops) – Area

In the development of area statistics for mixed cropping we follow the same procedure as for pure stand as described earlier in this chapter. What we are concerned with here is the apportionment of the areas among the various crops in the mixture. No single or uniform method can be suggested – the various possibilities are given below:

(a) Regard each constituent crop as occupying the whole area under the mixed crop. The snag here is that the total area computed this way exceeds the cropped area.

(b) Divide the area under the mixed crop on some concept of "proportion of area occupied". The method has the advantage that the area computed is equal to the cropped area. However, there are some difficulties in allocating the areas under the several crops. The method demands rather complex perception on the part of the observer. What he can assess on the ground is likely to be either the number of plants of the respective crops per unit area, or the proportion of the ground surface covered by the plants of the respective crops. The method is highly subjective and likely to produce satisfactory estimation only in the case of mixtures of similar crops at the same stage of growth or where fairly regular and systematic inter-cropping is done – otherwise judgement of the areas occupied by each crop is beyond the capacity of an average field officer.

(c) Convert the plants to their equivalent in pure stand. This requires an appraisal of density of plants in a reference pure stand and in the mixed stand. For each crop in a mixture, the ratio of the density of plants in the given field to the normal density found in pure stand conditions, multiplied by total area of the mixed crop gives the imputed area of the crop. The sum of the imputed areas of all crops in the mixture will generally differ from the actual area of the mixed crop.

(d) Regard the total area of the mixed crop as the area of the principal or predominant crop.

The criteria for selecting the principal crop will depend on the aims of the survey (e.g. crop occupying greatest area or giving greatest income). This method gives a greater area for some crops and less area for others. Another drawback is the non-availability of information on the areas of crops which are always the minor constituents.

In the author's view the most satisfactory method appears to be to give mixed areas separately from those of pure stand. Besides the usefulness in agriculture, this division makes estimation of production easy with crop cuts separately for pure stand and mixtures. For mixed crops, method (b) described earlier has been found convenient in many countries in allocating the areas. A table such as the one following would be of some use:

Area in ha/acre

Crops	Pure stand	Mixed crop	
		Principal crop	Subsidiary crop
Maize			
Cassava			
–			
–			
–			

5.7 Permanent Crops

For permanent crops, area component of production is meaningful only where planting is regular which is generally the case for large holdings/estates. In these cases the trees/palms are planted at fixed distances between each other and in almost straight lines. Further replanting is done so systematically that a constant ratio is maintained between bearing palms and total number of palms. Invariably such lands are cadastrally surveyed and the planted area figures available in the holdings/estates are satisfactory that it is not necessary to conduct any sampling surveys using physical measurements to improve on the estimates. It may be mentioned that in such holdings/estates, accurate production data are also available.

In small holdings, area component of production is not meaningful as the palm density per unit of land differs from holding to holding or even within a holding. The area component is therefore replaced by the number

Agricultural Production (Crops) – Area

of bearing palms. Data are generally collected for both the number of bearing palms and total number of palms.

For small holdings, data on the number of palms/trees can be collected by the reporting system. The methods for temporary crops can also be adopted for permanent crops.

The quality of the data collected can be improved by sample surveys with physical enumeration of bearing palms and non-bearing palms. An example of such a sample survey has already been given (Chapter 4) for coconuts in Sri Lanka. In this survey, enumeration of palms is done for the selected plots at the second stage of sampling.

CHAPTER 6

Agricultural Production (Crops) - Forecasting

6.1 Need

Forecast of production means anticipated knowledge of probable output. It can be defined as a statement of the most likely production of the crop assuming that factors such as weather, pests and diseases which may affect production between the time of forecast and final harvesting will be about the same as for an average year.

Forecast of production is needed for a variety of uses. Besides farmers and dealers, public and private organizations dealing with agriculture would be interested for providing the necessary storage and marketing facilities and for making available credit on the basis of crop prospects.

Governments require advance knowledge of crops as crops are an important factor in the national income. In countries which are not self sufficient in food, forecast of local production is needed to ascertain the quantities of foodstuffs to be imported. The Food and Agriculture Organization of the United Nations is regularly collecting advance information on crop prospects from the various countries in its early warning system so as to assist member countries for securing their deficit in food in time. In countries which produce cash crops and crops for export, forecasts are needed to make available to prospective buyers data on quantities for sale.

Further, production forecast is essential to all those who have a direct interest in the forecasting of prices of agricultural products.

6.2 The Two Components (Area and Yield)

Of the two components of crop production, area presents a comparatively simple problem. Ordinarily area is much more stable than yield. For permanent crops, changes in area are very slow. For temporary crops, the area planted can be ascertained at the time of planting, well ahead of harvest, to be used for early forecasts of production. The problem here is one of estimating rather than forecasting. Methods suggested in the previous chapter can be used for the area estimations.

In most developing countries, data on areas planted are collected at the beginning of the season for most of the important crops. An improvement in the forecasting can be made if harvested area is used instead of planted area particularly when yields are associated with harvested areas. The author has studied in Indonesia the close relationships between harvested area of wetland paddy for a four month period with: (a) the planted area for the four preceding months, and (b) the standing area at the end of the month immediately preceding the four month period. These relationships are found to provide a good basis for forecasting the harvested area for each four-month crop reporting period. When these average relationships over past years are measured by the regression method, the correlation coefficients are above $r=0.90$ for many years. These relationships are used in paddy forecasting in Indonesia on a regular basis.

When forecast of the area is needed before planting, this can be based on the economic relationship estimated from statistics for a series of years between prices of crops and other variables and the estimated planted area.

6.3 Crop Yields

The possibility of forecasting crop yields several years in advance would be of great value in the planning of agricultural production. However, the success of long-range forecasts is contingent not only on our knowledge of the weather factors determining yield, but also on our ability to predict the weather. As far as present knowledge goes no firm basis for reliable long-range weather forecasts has as yet been found.

The approach here is confined to methods of short-range forecasting which do not involve forecasting the weather, except by taking into account such inter-correlations as may exist between the weather observed up to the date of the forecast and subsequent weather conditions.

When considering crop production forecasting, the following three stages of the crop may be taken into account:

48   Agricultural Statistics: A Handbook for Developing Countries

   (i) Before planting;
   (ii) Growing crop;
   (iii) Just before harvest.

The last (stage iii) deals with the stage where crop cutting is done to estimate the yield and hence it is more an estimate of production than forecast and has been dealt with in earlier chapters.

For stage (i), the approach is based on enquiries from farmers of their intentions.

For stage (ii), there are three approaches:
   (a) Subjective method: experienced observers of crops estimate the yield on the basis of the growth;
   (b) Based on historical relationship between weather and yield to derive forecasts of yields (mainly using regression methods);
   (c) Objective method: measuring some characteristic of the crop in advance of the harvest which will provide some indication of the final yield.

From a practical point of view, forecasting from the growing crop is dealt with here at length.

6.3.1   Growing Crop: subjective method

In developed countries reports by local observers are the basis of forecasts of agricultural production. In developing countries these reports are generally made by the staff of the Agricultural Extension Services. Considerable progress has been made in developed countries in the collection and analysis of these reports.[1] These reports have been found useful in forecasting crop yields, especially late in the season after the reproductive stage of the plant's growth has advanced to the point where the fruit is easily visible. Reports made earlier in the season are less dependable due to the greater opportunity for extremes in subsequent weather to influence the final yield.

A condition report generally is a reliable indicator of the probable yield of crops whose vegetative appearance is highly correlated with yielding

---

[1] An excellent account is given in Chapter V of Fred H. Sanderson, *Methods of Crop Forecasting*, Harvard University Press, 1954.

capacity. Yields of crops with well defined fruiting habits and of those which are subject to a critical period of relatively short duration (during which the plant is most sensitive to weather influences) are generally more accurately forecast by condition reports than those of crops where fruiting is continuous (like cotton). Condition reports, however, are not reliable indicators of prospective yield where there is little correlation between the appearance of the vegetation above ground and the economic yield of the crop, as in the case of roots and tuber crops.

Condition reports are made by experienced observers in terms of percentages of the full or normal crop. However, it is unlikely that observers can identify more than a small number of sub-categories between crop failure and crop perfection. A working arrangement could be to report the percentages in multiples of 10. The full or normal crop can be taken as one that is healthy and undamaged by drought, pests or accidents and which shows the state of growth which is reasonable under such conditions.

One of the chief objections to the use of condition reports is the fact that the individual reports themselves as well as the method of sampling are subject to a systematic bias which is not reduced merely by increasing the size of the sample. But if a stable relationship exists between reported condition and final yield, and if accurate and unbiased estimates of final yield are available for a number of years, the bias can be eliminated by fitting a regression line to the scatter of reported conditions, versus final yield. If, however, there is doubt as to the stability of the condition-yield relationship, or the freedom from bias of the final yield estimates, the subjective method is of limited value.

Where there are historical data which will provide a guide to the usual relationship between condition reports and final yields, a simple way to find the likely yield is from a graph. When a graph is drawn of condition reports against final yields which indicates a relationship, the probable yield corresponding to the condition figure for the current year can be read off. Such relationships have a tendency to change with changing technical conditions and this has to be watched.

Though this is a workable method, cases arise where there is a major deviation between condition reports and final yields. This may be due to the variability of pest damage or rainfall at a critical stage in the growth. This can be overcome by making the condition reports still later in the season to cover the critical period, provided the forecast is timely. Or as an alternative, adjustments can be made by introducing additional variables into a multiple regression analysis with reported condition and final yield. The important additional variables are time trend, pests, and weather factors. It is perhaps worth pursuing such adjustments because condition reporting system is likely to be in use for many years as countries would take a long time to develop objective methods for forecasting yields.

6.3.2 Growing Crop: second method

If we accept that weather in the growing season is highly influential on the crop growth then from historical relationship and current observations of meteorological data, a good forecasting system should be derivable. Although a large amount of the work has been done (mainly in developed countries), this method has not made an impact on the forecasting system. The problems to be faced in this method are:
(a) Sensitivity of the crop to conditions in specific short period of its growing phases and meteorological records for these periods may not be easy to assemble;
(b) Important weather factors may occur late in the season.

In this method, various weather relationships and yield are explored using regression analysis. The procedure of plotting the yield against the most important variable and drawing a line of best fit, then plotting the deviations from this line against the second factor and so on, may give ideas.

Although this method is mentioned as one of three possible methods for dealing with the growing crop, experience has shown that weather should be treated as providing a modifying factor for the other two methods rather than as the main indicator.

6.3.3 Growing Crop: objective method

In this method efforts are focussed on the possibilities of forecasting yields from measurements of plant characteristics. Some of these are direct components of yield. For example, in cereals, the number of plants per unit of land, the number of heads per plant, the number of kernels per

Agricultural Production (Crops) − Forecasting

head and the weight per kernel are direct components of yield. Other characteristics, though not direct components of yield, are correlated with yield like the thickness and height of stalks. We shall, however, confine our attention to measurements of some direct components namely the progress of particular fruits through the season, in relation to the stage of maturity reached at the time of observation.

As in the case of condition reporting, in objective methods too it is preferable to take the measurements in the later part of the season after the fruit or other commercial product is set. Earlier in the season, before the individual fruits are set, it is more difficult to find reliable indicators of the yield.

In this method, namely forecasting yield in late season from measurements of fruits, there are three elements to be considered:

(i) The number of fruits at the time of observation;
(ii) The number that is likely to survive for harvesting;
(iii) The weight of the fruits.

In recording the number of fruits at the time of observation, it is necessary to classify them into maturity classes as the probability of survival will be different for the different classes. This requires a definition of maturity class. It is easy for cotton as it has clearly demarcated stages. It is difficult for maize. Hence maturity classes should be defined for each crop. Consultations between the statistical staff and agricultural botanists would be needed to establish standards for classification for the several crops.

The next step is to estimate the probability of survival till harvest for each maturity class. These probabilities are estimated from observations made in previous surveys by noting the survival of tagged fruit on sample plots. If N_i is the number of fruits in maturity class i and p_i is the probability of survival in that class then the forecast of the total number of fruits for harvesting is the sum of $p_i N_i$ for all the maturity classes.

To complete the forecast the weight of fruit per unit at harvest is needed. This can be obtained from historical averages. For crops for which final yields are estimated by crop cutting, these averages would be readily available. Otherwise for a start arrangements could be made with organizations like the cooperatives to weigh some of the harvested crops for estimating these averages. These, however, would require refinements with the passage of time.

Though the methodology of the objective method looks simple, it is no easy task for application in the field on a large scale in developing countries. Research Stations of the Agricultural Departments/Ministries in these countries could initially undertake the study of the relationships between plant characteristics and yields particularly those relating to direct components (fruits, grains) for the selected crops. When these studies reveal a stable relationship for a crop, the measurements of the crop in the course of growth can be added to the country's routine field work for collection of agricultural statistics. It is perhaps wise to restrict the number of crops to one or two so as to evolve suitable working techniques and at the same time to lessen the burden on the statistical field staff.

For the use of regression methods, objective methods involving measurement of crop characteristics and other methods for forecasting yields, historical data would be needed. The Research Stations of the Agricultural Departments/Ministries in developing countries would be a good source for such data though sometimes the data may not cover the whole country. However, these data could make a starting point for evolving suitable methods which could later be modified with better data (quality and coverage) and more experience. The Research Stations could be made use of for collecting data which otherwise would be difficult to collect (e.g. data necessary to set up maturity categories).

The Meteorological Departments would be the source for meteorological data.

CHAPTER 7

Agricultural Production - Livestock and Livestock Products

7.1 Introduction

Collection of livestock statistics in developing countries is, in many ways, a more arduous task for a statistician than that of crops. In the case of crops one deals with land (immovable) while for livestock, the animals in most cases move about making difficult collection of data. Large farms cause no problem as the farm owners/holders maintain good statistics. It is the small holder who makes the collection difficult and expensive. Further it is not always that he maintains livestock for its economic use; sometimes the economic use is secondary to the social value of the stock, rendering the collection of statistics unimportant for economic development.

Collection of livestock statistics as for other agricultural statistics is done in three ways:
  (i)   Administrative reporting;
  (ii) Sample surveys;
  (iii) Censuses.

7.2 Livestock Inventory

For livestock numbers, the real starting point should be a census. It is in a census that basic data such as numbers, age and sex could be determined to the lowest administrative or geographical unit needed. In between the censuses, needed data have to be collected by administrative reporting or by sample surveys.

It has been found by experience that basic data like livestock numbers can conveniently be collected by administrative reporting despite some shortcomings. Sometimes resort is made to sampling. These data are generally collected annually. Other data like livestock structure can generally be collected by sample surveys conducted about once in three years. These surveys would need specially trained officers and would be difficult for administrative reporting.

In the collection of current statistics for crops the items are few, like area/number of trees, yield and production. In livestock, however, there are many items and therefore it is necessary to select the items for data collection beforehand. Some items[1] for data collection are listed below to indicate the variety:

(1) Information on the structure of livestock holding;
(2) Information on the composition and on the structure of livestock;
(3) Study of yearly flows (sales and losses);
(4) Study of information useful for agricultural advisory purposes;
(5) Study of data needed for the calculation of zootechnical parameters (e.g. fertility rates);
(6) Data needed for the calculations of economic parameters (e.g. percentage of calves retained for the breeding stock within a generation);
(7) Data necessary for short and medium term forecasts;
(8) Information needed for agricultural accounts (inter-regional trade, increase in livestock numbers).

Livestock population is subject to marked seasonal fluctuations resulting in maximum and minimum numbers within the course of the year. Hence in annual collection of livestock numbers, two enquiries are suggested. It would be desirable to investigate more frequently the number of those species considered to be of special economic interest and of greatest seasonal variation, such as pigs and poultry.

---

[1] It may not be feasible to collect data on all the items mentioned but these are given for the sake of completeness.

7.2.1 Administrative reporting

It is convenient to start building current statistics on livestock inventory the year after an agricultural/livestock census which provides the bench mark for basic data. The smallest reporting unit should be for which a veterinary or livestock or an agricultural extension officer is available whose duties include collection of data on livestock. When collection commences the year after the census, the officer has before him the census data. He has merely to note down the changes during the year to give the data for the year. This procedure is repeated till the next census.

In this method of collection, items should be restricted to number, age and sex. For age FAO's classification recommended for the World Census of Agriculture would give guidance, e.g.

Horses	under 3 years
	3 years and over
Camels	under 4 years
	4 years and over
Cattle	under 2 years
	2 years and over
Buffaloes	under 3 years
	3 years and over
Sheep and goats	under 1 year
	1 year and over
Pigs	under 6 months
	6 months and over

Besides the traditional reporting system there are other administrative reportings. Where livestock is taxed, taxation returns would give some data. Other sources would be book-keeping documents of cooperatives/ state farms, pasture areas and drinking places (particularly for nomadic livestock).

Estimates from administrative reporting could always be improved by sample surveys at suitable intervals from which correction factors could be evolved.

56 *Agricultural Statistics: A Handbook for Developing Countries*

7.2.2 Sample surveys

7.2.2.1 Frame

The frame of sample surveys for livestock poses some special problems unlike for crops where some frame is generally available of cropped or land areas. For livestock, normally a frame is available after an agricultural or livestock census. If sample surveys are to be conducted not long after a census, the census frame could be used after updating it. Otherwise, instead of preparing a new frame which involves heavy expenditure, staff and time, every effort should be made to make use of records available (with administration, Department of Agriculture/Livestock/Veterinary Services). Lists prepared from these records should be brought up-to-date before use. There are cases of holdings with livestock but with no cropping of land (called landless holdings) and care should be taken not to exclude them.

Different possibilities for constructing a frame are given below:[1]

(1) List of holdings and household;
(2) Taxation registers;
(3) Book-keeping documents of cooperatives/state farms;
(4) Census of drinking places;
(5) Pasture areas;
(6) Aerial photography taken at the beginning or at the end of the day in order to avoid the problems raised by animals that seek the shade;
(7) Area sampling;
(8) Compulsory insurance schemes;
(9) Veterinary lists;
(10) Comparisons with deliveries or receipts of hides and skins.

7.2.2.2 Sample design and enquiries

Stratification can be by agricultural or administrative or geographical regions. Primary sampling units can be villages or communes and the

---

[1] Some of these may not be feasible mainly due to high costs but are mentioned for the sake of completeness.

Agricultural Production – Livestock and Livestock Products

second stage units livestock holdings.

Calves may be investigated individually and the product related to the mothers; the destination of calves and young cattle can be asked. Sheep must be surveyed by "lots" and fertility studied globally for each sheep holding. A survey of goats can be carried out on the basis proposed for cattle and sheep.

As regards pigs, due to their short production cycle and their rigid economic cycle of 36 months, two methods may be used:
- (a) Surveys of pig number every 3 or 4 months (which permit the evaluation of the parameters needed for forecasts);
- (b) Continuous sample surveys of boar owners.

For hen numbers, survey can be carried out by repeated visits with a fixed sample, or with a rotation sample or a sample with partial replacement for obtaining monthly estimates. The reference period can be the previous day or the whole previous week.

In livestock surveys, generally, more frequent visits by enumerators are needed than for crop surveys where one or two visits would suffice. Farmers tend to forget certain items over a period of time – hence the need to contact them at the appropriate times. Every new visit should record the changes like additions to the herd, births and purchases and numbers leaving by slaughters, sales and deaths.

7.2.3 Nomadic livestock

Though there are many methods to estimate nomadic livestock in theory, only a few have been tried in the field. An example[1] is given here from Somalia of estimating nomadic livestock population using sampling, taking watering points (for drinking water) for the frame to illustrate two methods tried in the field. The survey was conducted during the dry season because in the wet season livestock could use other sources of water. The frame consists of the watering points in the domain of study, which is reasonably simple to prepare. The primary sampling units are the watering points and the second stage units are the households.

---

[1] Nadarajah, *Report on Statistics to Somali Democratic Republic*, 1976.

58   Agricultural Statistics: A Handbook for Developing Countries

Two methods have been tried in Somalia namely:
(1) "one day" model;
(2) "extended enumeration" model (14 days).

7.2.3.1   *"One day" model*

Enumeration takes place for 24 hours from 6 a.m. one day to 6 a.m. the following day at the selected watering points by interviewing the households who bring the livestock to the watering point concerned.

This method assumes the following background local conditions. The tradition bound watering practices of nomadic households assume a highly stable and uniform pattern, especially during the dry months of the year. Individual nomadic household conforms to a routine practice of watering each of its livestock types at *regular intervals* during the dry season. This means that watering interval of each livestock type[1] of an individual household is generally fixed and if it does vary, it does so within narrow limits due to the action of chance causes only. Further nomads exercise no conscious preference or non-preference as to choose any particular day/days of the week to water their animals.

Precautions are taken in this method as well as in the other method:
(a) For eliminating at the interview (at the watering points) those from settled households;
(b) Against obtaining information from more than one member of the same household;
(c) Against steps to be taken from enumerating the same household from more than one watering point.

The estimated population for one livestock type is given by:

$$\hat{Y} = \frac{N}{n} \sum_{i=1}^{n} \sum_{j=1}^{m_i} w_{ij}\, y_{ij} \qquad [2]$$

---

[1] Sheep/goats, camels, cattle.

[2] In practice the formula becomes simpler as the watering interval tends to remain uniform for each livestock within a district.

with the usual notations where w_{ij} is the watering interval (number of days) for that livestock type in the jth surveyed household of the ith sample watering point.

7.2.3.2 *"Extended enumeration" model*

Enumeration commences from 6 a.m. on a particular day and continues for 14 days.

The assumptions in this method are:
(a) All members of nomadic households tending livestock[1] water their herds, in each case, at intervals never or very rarely exceeding 14 days;
(b) Every household in the population enjoys an almost certain chance of being contacted at least once, at one or another watering point during the 14-day enumeration.

To achieve the correct objectives, the questionnaires are prepared to meet these requirements. Households are treated as composed of three mutually exclusive groups (or subsets of households):[2]

Subset I. Owning sheep/goats with or without other types of animals (identifying livestock type common to all households in this subset-sheep/goats);

Subset II. Rearing camels but no sheep/goats with or without cattle (identifying livestock type common to all households — camels);

Subset III. Rear cattle only.

The estimated population for one livestock type is given by:

$$\hat{Y} = \frac{N}{n} \sum_{i=1}^{n} \sum_{j=1}^{m_i} y_{ij}$$

with the usual notation.

---

[1] Sheep/goats, camels, cattle.

[2] The situation holds for the other model as well.

7.3 Livestock Products

Livestock products cover a wide variety of items among which the important ones are meat, milk and milk products, and eggs.

Some of these items are covered by regular administrative reportings[1] such as returns on the number and variety of animals slaughtered at slaughter houses. Such data would only cover part of the total slaughterings. These data, however, would form the core of basic available data. As coverage in these cases is limited, sample surveys are needed for the rest of the population. Some of these surveys directly estimate production while in other cases indirectly. In the latter case quantities available for consumption are estimated and from these estimates, production is estimated.

7.3.1 Meat production

For meat statistics to be useful particularly for national planning and international comparability, a number of requirements have to be fulfilled. There should be complete coverage for all slaughterings taking place in public slaughter houses, and on farms and in villages. They should distinguish between the slaughterings of indigenous animals, imported animals and animals exported for slaughter. Compilation of statistics should, as far as possible, meet with concepts and definitions of basic terms in international use. Meat statistics should provide data giving an indication of the end-use of livestock products such as local consumption, export or utilization by processing industries.[2] Further they should provide average liveweight and carcass weight of slaughtered animals by species, differentiating between improved and unimproved animals and between important livestock groups based on age and sex and data on by-product of slaughterings (offals, slaughterfats). In order to satisfy these requirements, meat statistics have to be based on adequate sources of information and the application of objective methods and procedures for data collection.

---

[1] Generally made by veterinary officers.

[2] Most of these data would be needed for the compilation of the annual food supply/utilization accounts of the country.

7.3.1.1 Concepts and definitions

FAO has recommended the two concepts "gross indigenous production" and "meat available for consumption" for use for the countries.

Gross indigenous production is defined as:

Meat for all livestock slaughterings in the country irrespective of their origin;
+ the meat equivalent of live animals exported;
− the meat equivalent of live animals imported.

Meat available for consumption is defined as:

Meat from all livestock slaughterings in the country irrespective of their origin;
+ imported meat and the meat equivalent of derived products;
− exported meat and the meat equivalent of derived products;
± change in stocks of meat and the meat equivalent of derived products.

In meat statistics, one comes across terms like live weight, killed weight, dressed carcass weight, offals and slaughter fats. These may be described as below:

Live weight of animals intended for slaughter is the weight taken immediately before slaughter.

Killed weight is the gross weight of the carcass before dressing including the hide or skin, head, feet and internal organs, but excluding the blood.

Dressed carcass weight. There are variations between animals but generally can be stated as the weight of the carcass after removal of:
(a) the hide;
(b) the head;
(c) the tail;
(d) the offals;
(e) the slaughter fats.

Offals consist of tongue, brains, heart, liver, lungs, throat, spleen, diaphragm.

Slaughter fats. Edible fats include loose fats (in the abdomen or

thoracic cavities), kidney fats and back fats (such as lard and flore fats). Inedible fats include fats from discarded animals or carcass guts, fats recovered from sweepings, hide trimmings etc.
Countries should as far as possible adhere to these concepts.

7.3.1.2 Data collection: reporting system and sample surveys

The production recorded in the reporting system refers only to livestock slaughterings carried out at controlled public slaughter houses and meat-packing plants, excluding the production from farm slaughterings which are not generally subject to veterinary inspection or fiscal control for taxation purposes. The data from slaughter houses are collected either daily or weekly or monthly depending on the importance of the slaughter houses and returns are sent to the Veterinary or Local Authority. In all cases the number of each variety of animal is reported while in a few cases, in addition, the weights of animals. Where weights are not recorded, the meat production is estimated as the product of the number of animals and the average carcass weight of the animal.

Uncontrolled slaughtering may be estimated by using modern statistical techniques suitable to the national conditions. Food consumption surveys based on economic classes of families and sub-regions could provide information on the meat available for consumption. From consumption estimates, production can be estimated.

Surveys to estimate the number and structure of livestock can also be made use of for data on uncontrolled slaughtering.

In the absence of adequate sample surveys, the production of meat may roughly be estimated on the number of hides and skins collected and by using average slaughter rates based on national herd models.

In view of what has been stated, it is suggested that countries:

(a) Make use of the reporting system for controlled slaughterings by collecting data on number of animals and other important items (e.g. carcass weight);
(b) Make use of surveys conducted for food consumption or number and structure of livestock for uncontrolled slaughterings, for estimating meat production.

7.3.1.3 Poultry

Estimating production of chicken for meat and for eggs has to be treated separately as the cycle of production is very short compared to other livestock.

Taking egg production first, production can be estimated on the basis of number of hens and of the number of eggs laid per hen. Hen numbers can be assessed by means of a sample survey on the basis of a census. The survey can be carried out by repeated visits with a fixed sample, or with a rotation sample or a sample with a partial replacement, for obtaining monthly estimates. The reference period can be the previous day or the whole previous week. On a sub-sample basis, number of eggs per hen, weight and eggs and losses can be determined.

Sometimes egg production is determined directly from a sample of producers, where complete lists of producers are available.

In countries where incubators are used for hatching eggs, an alternative method[1] is from the number of day-old chicks. It is necessary to make a distinction between chicks intended for meat and for egg production. Further, data should be collected on losses at each stage, the age of commencement of lay, the age at cullings and on natural incubation as well. Such surveys can be carried out by postal questionnaires together with interviews in case of non-response.

For determining the production of chicken for meat, it is possible to follow the production of day-old chicks and the production of slaughter houses, and to carry out sample surveys similar to those for hens laying eggs. The reference period can be one week or one month and producers can be stratified by volume of production. Monthly visits or collections are not necessary; quarterly visits/collections should suffice.

7.3.2 Milk and milk products

Statistics of milk and milk products are known to present the most intricate problems since they are produced on farms and in villages where

---

[1] This method would cover the sector using incubators – for the other sector one of the other methods described in the section has to be used.

they are consumed in substantial part without ever reaching markets, processing plants, or inspection centres that would enable reliable information to be collected. Whole milk production, whole milk utilization, butter and cheese, and other dairy products (cream, condensed and evaporated milk, dried milk, casein, skim milk, etc.) would come under this heading.

Generally milk production would refer to milk actually milked excluding that sucked by calves, lambs and kids. Production should include:
  (a) Milk delivered to commercial dairy plants or purchasing centres;
  (b) Quantities used by producers at places of milk production;
  (c) Milk sold directly to consumers;
  (d) Milk fed to livestock;
  (e) Farm losses.

For complete coverage, production should include production at non-farm households located in urban areas. It is well known that a significant number of urban households in developing countries maintain milk animals (cows, buffaloes, goats) for direct consumption. Coverage should also be extended to cattle posts located at great distances from settled areas where the milk produced is wholly utilized by the herders.

Data on milk utilization are required to ascertain quantities utilized for human consumption and for feeding livestock. These data can be used for mutual checking of estimates on the sides of production and consumption. These data are also needed for calculation of index numbers of food and agricultural production, economic accounts for agriculture and food supply/utilization accounts. The main items of utilization could be as follows:

I. Utilization at place of production (i.e. on farms)	II. Utilization at dairy plants
(a) for farm household consumption	(a) for fresh consumption (raw, standardized, pasteurized)
(b) for direct sale to consumers	(b) for manufacture of cream, butter, cheese, condensed and evaporated milk, dried milk
(c) for processing of cream, butter, cheese-on farms	(c) manufacture of by-products (casein)

Agricultural Production – Livestock and Livestock Products

(d) for livestock feed
(e) farm losses
(f) for deliveries (sales to dairy plants, milk purchasing centres)

(d) factory waste

In most developing countries, available milk production statistics relate to item I(f) only. Data for items II(a) to II(d) are available with the relevant organizations concerned.

Butter and cheese, in most developing countries, are produced partly on farms and partly in dairy plants. In some countries where dairy industry is not, as yet, reasonably developed subsistence production of butter forms bulk of total production. On the other hand, farm cheese production by traditional methods, is not considered significant in these countries. Though farm butter production is important in many countries, data available are not satisfactory. As to factory production of butter and cheese, the relevant data are obtainable from the records of the plants.

The production of other dairy products mostly takes place in dairy plants as such production data of each are available from the records maintained in the plants.

What has been described relates to definitions and concepts and coverages of the investigations. What follows will relate to sources of information and methods of collecting data.

7.3.2.1 Sources of information and methods of collecting data

Production is determined either by direct methods or by indirect methods as below:

Direct methods
(1) As the product of the number of producing animals (buffaloes, cows, goats) and the average yield per animal.
(2) As the sum of:
 (a) sales or deliveries to dairy plants
 (b) direct sales to consumers
 (c) estimate of milk used on the holdings

Either method can be used. Sometimes a combination of the two methods is used.

Indirect methods
  (1) As the product of estimated *per caput* milk consumption and the total number of persons.
  (2) As the sum of quantities of milk used for:
     (a) fresh consumption
     (b) manufacture of milk products
     (c) livestock feed
     (d) losses (at the producers and dairy plants)

Data for (b) and (c) are either collected directly or estimated by converting the various dairy products used into their milk equivalents by suitable conversion factors.

For direct method (1), it is necessary to have estimates of the number of producing animals and the average yield per animal. The number of producing animals is estimated on the basis of the last census corrected by annual reportings or surveys held on the number and structure of livestock. The average milk yield is best obtained through a sample of dairy farms. In the absence of current sample surveys, state livestock farms in most countries are able to provide reliable estimates of average yields.

For direct method (2), figures for item (a) are the only ones known with accuracy in most countries – (b) and (c) have to be estimated from sample surveys.

For estimating production, generally direct methods are preferable. It is, however, not always possible to do so. Hence some countries resort to indirect methods. The indirect methods refer broadly to estimation of consumption of whole milk, milk products, and losses. Many countries conduct regular consumption surveys and therefore these surveys can be made use of for estimation of consumption of milk and milk products. From consumption figures with suitable modifications production can be estimated.

7.3.2.2 *Sample surveys*

Any livestock products survey should be preceded by a census or survey with accurate data on the number and structure of livestock (sex, age, race, size of stock-holdings etc.). This applies to milk and milk products as well.

In countries without sufficiently widespread milk control or systematic deliveries to cooperatives or sales organizations, sample surveys can be carried out having in view two objectives:

(a) Collection of data on milk production classified according to the following criteria: age, race, size of livestock holdings;
(b) Collection of data on the utilization of milk produced on farms.

7.3.2.3 Sample design

Stratification by agricultural/geographical/administrative regions.
Primary or first stage sampling units: villages;
Second stage units: livestock holdings/farms.

To provide for seasonal variations of production and for the average duration of the lactation period, survey should cover about 14 months for cows. The reference period could be a day and frequency of visits – once a month. To check on production figures given by producers to enumerators who collect the data, cows are selected at random in the sampled holdings and milk weighed. Analysis of fat content can also be made.

The sample design can be varied where milk production is on some organized basis. In this case the frame can be cooperatives/sale organizations or processing plants. These can be stratified according to size of these organizations (i.e. according to the quantity of milk handled) and a random sample selected and data collected. Suitable correction factors can be evolved by actual measurements on a sub-sampling basis.

The milk production of goats and ewes can be investigated in the same way.

7.3.3 Food consumption surveys

As data for estimating livestock products are sometimes secured from food consumption surveys, the subject is briefly described here. These surveys and Food Balance Sheets are the two sources from which information could be obtained on food supply and consumption. For these surveys to be useful, they should be conducted on a large scale with regional breakdowns and with information by population groups.

The sample design for these surveys is similar to other surveys in agricultural statistics. In addition, it would be necessary to stratify into urban and rural areas.

Time problem is an important one in these surveys. It is important (i) from the viewpoint of the time coverage (period to be covered by the survey) and (ii) from the viewpoint of the length of the reporting period. Time coverage should be such as to allow for seasonal variations. Regarding the length of the reporting period, it depends on the durability of the goods. Perishable goods might be bought frequently and also used immediately. Other goods may be acquired at longer intervals. As such, stratification according to the durability of the goods might be considered. It is, however, not wise to use too long reporting periods because of the end-effects in interview surveys and fatigue in households who keep accounts. Furthermore, long book-keeping is likely to influence consumption pattern of a household.

If cooperation of the household could be secured, goods should be weighed. As data have to be mostly secured from housewives, female enumerators have been found useful.

Though random sampling is preferable, some countries prefer purposive selection due to considerations of cost and the voluntary participation in this kind of survey.

CHAPTER 8

Index Numbers of Agricultural Production

8.1 General Remarks

Index generally means a pointer showing measurements. In our case, it measures changes in production from year to year. For measuring yearly changes a base period is needed to measure from. The base period should be more than one year to absorb any abnormal fluctuations in the productions which may take place if one year is taken.[1]

Merely adding the volumes of production of the different categories of agricultural commodities to measure production is not meaningful. It is, however, meaningful when the product of the volume of production of the various commodities by their prices is considered. Here the sum of the products means the value of total agricultural production (using prices as weights).

Thus in any work on index numbers of agricultural production, it is necessary to consider (i) the volume of production for the year concerned and that for the base period and (ii) some form of weighting these volumes. Weights generally used in the construction of national index numbers are the average producer prices of the commodities concerned for the base period. If the base period is taken as five years then the weights are the average producer prices of the five years.

8.2 Concept of Production

One concept is that currently used in many countries and the FAO, namely the aggregate volume of agricultural production flowing to all

[1] FAO uses at present 1961–65 as base period.

sectors other than agriculture itself. The other concept of production is that used by United Nations for computing index numbers of industrial production called "value-added" type.

In the first concept, agricultural commodities used in the agricultural production process are deducted from total production. Those include seed and livestock feed (i.e. agricultural products fed as such and semi-processed feed such as oilcakes and bran). Index numbers compiled using this concept may be called "supply-oriented type": these measure the rate of growth of the outgoing supplies of agricultural commodities to meet demand. This is the relevant rate of growth in comparing the rate of increase in the requirement of farm products for all uses other than intermediate consumption in agriculture. Such measure is useful for the various analyses of the agricultural situation and problems.

The second concept is the one used in economic accounts of agriculture. For example, to estimate the gross national product, one must deduct from the measure of aggregate production according to the first concept, purchases from the non-farm sector of materials and services to be used in agricultural production. This is the "value-added" concept. The "value-added" index numbers are important as they reflect the growth of the domestic product derived from agriculture within the larger context of overall economic growth.

8.3 Weights

Production of agricultural commodities can be aggregated in many ways depending on the weighting system used. The weight commonly used for a commodity is its average producer price for the base period. In actual practice, it is convenient to use the average producer price of the commodity in relation to the corresponding price of wheat (by weight and not by volume) expressed as a percentage. Such a number is known as the price relative of the commodity. In countries where wheat is not grown, the prices can be related to price of rice.

8.4 Formula Used

There are many formulae used to measure or calculate the index of production. The most common one is by Laspeyre. The basic assumption

Index Numbers of Agricultural Production

here is that price relatives fixed for the base period do not change for the whole period for which the index is computed.

Laspeyre's formula for the index of a year (t) is given by

$$\frac{\Sigma p_0 q_t}{\Sigma p_0 q_0}$$

given as a percentage of $\Sigma p_0 q_0$, where p_0's are the weights of the different commodities for the base period (i.e. the average price relatives of the base period), q_0's are the average productions of the different commodities for the base period, and q_t's are the productions of the different corresponding commodities for the year (t) for which the index is required.

In simple language, here the index measures the change annually of the value of total agricultural production at constant prices (namely the average national producer prices of the base period).

8.5 Indices Used

Among the various index numbers in use in agricultural production, the prominent ones are:
- Index of agricultural production;
- Index of food production;
- Index of *per caput* agricultural production;
- Index of *per caput* food production.

Indices for food production cover the following commodity groups: Cereals, starchy roots, sugar, pulses, edible oil crops, nuts, fruit, vegetables, wine, cocoa, tea, coffee, and livestock and livestock products.

Indices for agricultural production cover all the items included in food production, and, in addition: industrial oil seed, tobacco, fibres (vegetable and animal) and rubber.

Meat production for purposes of index numbers is in terms of live weight. If national production figures are available in carcass weight, these have to be converted to live weight. The conversion factors are those pertaining to the base period.

72 Agricultural Statistics: A Handbook for Developing Countries

The national index numbers of agricultural and food production over a period of years would give the trends in these productions. The trends, however, would be different when changes in population are taken into account. For example, in a country where the index of agricultural production for a particular year was found to be 141, the index of *per caput* agricultural production was found to be only 102.8[1] the reason being that the population for the year was 137.2% of that of the base period. Thus *per caput* index numbers are useful as they throw additional light on the progress of production. The index of *per caput* food production is very useful in countries which are not self sufficient in food as this would measure the progress achieved and the gap to be filled, taking the population into account.

8.6 Example

An example here illustrates how the various index numbers are calculated. For simplicity the commodities are restricted to five out of which four are food commodities and one (rubber) non-food. The base period is 1961–65 and index numbers are worked out for year 1974. The data given refer to an Asian country which appear in *FAO Production Yearbook of 1974*, suitably corrected so that production figures are according to the first concept mentioned in Section 8.2.

		Rice	Maize	Cassava	Beef and Veal	Rubber
Production in 1000 metric tons	Base[2] (q_0)	8000	2700	11 800	262	700
	1974 (q_t)	14 500	2650	9400	270	890
Price relatives	(p_0)	114	77	29	485	529

[1] $\dfrac{141.0}{137.2} \times 100 = 102.8$.

[2] Average for 5 years (196 in 1965).

In using Laspeyre's formula

$$\frac{\Sigma p_0 q_t}{\Sigma p_0 q_0}$$

for agricultural production, the index for 1974 is given by

$$\frac{114 \times 14500 + 77 \times 2650 + 29 \times 9400 + 485 \times 270 + 529 \times 890}{114 \times 8000 + 77 \times 2700 + 29 \times 11800 + 485 \times 262 + 529 \times 700}$$

expressed as a percentage which is 139.4.

The index of food production for 1974 is given by

$$\frac{114 \times 14500 + 77 \times 2650 + 29 \times 9400 + 485 \times 270}{114 \times 8000 + 77 \times 2700 + 29 \times 11800 + 485 \times 262}$$

expressed as a percentage, which is 142.3.

For calculating the *per caput* index numbers, it is given that the 1974 population is 135.7 taking 100 for that of the base period 1961–65.[1]

Thus the *per caput* index numbers are for:

(a) Agricultural production $\frac{139.4}{135.7}$ as a percentage, which is 102.7;

(b) Food production $\frac{142.3}{135.7}$ as a percentage, which is 104.9.

---

[1] Average for 5 years (1961–65).

CHAPTER 9

Supply / Utilization accounts (Food Balance Sheets)

9.1 Introduction

These accounts are prepared by countries annually — mostly at the national level. A calendar year has been found convenient. For purposes of comparison, the accounts prepared for an earlier period of three years (say 1963-65 or 1964-66) would be found useful as a base. As the name indicates, the balance sheets would deal with supply on the one hand and utilization on the other for the whole country. Supply would come from production and imports adjusted for change of stocks. While utilization would deal with exports, feed, seed, manufacture and waste; and what is left of utilization is food for human beings, i.e. food supplies available for human consumption at the retail level (as the food leaves the retail shop or otherwise enters the household). Food may be consumed directly like potatoes, cassava, bananas[1] ... or after processing like rice, wheat, flour....[2] The total food consumed for the whole country is then divided by the total population (for the mid-period) to give the *per caput* consumption by kg/year, g/day, calorie/day, protein/day, and fat/day.

9.2 Purpose and Use

The purpose is as the title indicates to balance the food supply and utilization (at the national level). The balance sheets indicate the magnitude of the import gaps, quantities available for export and change

---

[1] Called primary products.

[2] Called derived products.

Supply/Utilization accounts (Food Balance Sheets)

in stock position. They further give the levels of *per caput* consumption of calorie, protein and fat intake at the national level. If and when pockets or regions of low food intake have to be identified then the accounts should be prepared at provincial or district levels. The relative importance of the different foods in the dietary pattern and contribution of the different foods to the total *per caput* supplies of calories and proteins can be seen from these accounts.

The availability of these accounts annually is of great importance to a nation as it will enable the policy makers and planners to remedy the nutritional inadequacy of the available diet at the national level. In the preparation of these accounts, gaps and deficiencies in agricultural data collection are brought to light. For example, some items of food considered unimportant from the point of view of data collection may be found to be important for calorie/fat/protein intake. The quality of some of the statistics collected may not be found dependable when checked with related supplementary information from food consumption and dietary surveys. The accounts may also reveal other gaps such as inadequacy of data on stocks.

9.3 Commodity Coverage

The various commodities covered in the food balance sheets are given below, some are single items while others are grouped:

I. *Cereals*
   (1) Wheat
   (2) Rice
   (3) Coarse Grains
       Maize
       Barley
       Oats
       Millet and sorghum
       Rye
       Others n.e.s.

n.e.s. Not expressly stated.

II. *Starchy Roots and Tubers*
  (1) Potatoes
  (2) Sweet potatoes
  (3) Cassava
  (4) Yams
  (5) Others n.e.s.

III. *Sugar*
  (1) Sugar, centrifugal
  (2) Sugar, non-centrifugal
  (3) Syrups
  (4) Others n.e.s.

IV. *Pulses, Nuts and Oilseeds*
  (1) Pulses
  (2) Nuts and kernels
  (3) Oilseeds

V. *Vegetables*

VI. *Fruit*
  (1) Citrus fruit
      Oranges and tangerines
      Lemons and limes
      Others
  (2) Bananas
  (3) Other fresh fruit
  (4) Dried fruit

VII. *Meat (carcass weight)*
  (1) Beef and veal (including buffalo)
  (2) Mutton, lamb and goat-meat
  (3) Pigmeat
  (4) Poultry meat
  (5) Other meat n.e.s.
  (6) Offal

n.e.s. Not expressly stated.

Supply/Utilization accounts (Food Balance Sheets)

VIII. *Eggs (Hen and Duck)*

IX. *Fish*
 (1) Fresh water
 (2) Marine

X. *Milk and Milk Products*
 (1) Whole milk
 (2) Skimmed milk
 (3) Dried milk
 (4) Cheese

XI. *Fats and Oils*
 (1) Butter (including ghee)
 (2) Vegetable oils
 (3) Animal fats (including marine oils)

XII. *Other Commodities*
 (1) Spices
 (2) Cocoa
 (3) Non-alcoholic beverages
   Coffee
   Tea
   Soft beverages
 (4) Alcoholic beverages
 (5) Other food n.e.s.

9.4 Concepts and Definitions

Production

For primary items, production relates to the total domestic production. For primary crops, production is reported at the farm level (harvested production). Meat production is that of dressed carcass weight of animals. Fish production is that of liveweight (i.e. the actual ex-water weight of the catch at the time of capture).

Production of processed commodities relates to the total output of the commodity at the manufacture level.

In the food balance sheets (format given at the end of the chapter), production has two columns — column (1) input, and column (2) output. Primary products such as paddy and potatoes are entered in column (2) as there is no input in these cases. When these products are consumed directly like potatoes, there are no more entries for this product in the balance sheets. When the product is not consumed directly like paddy, after its utilization in paddy form, the balance is entered a second time (in the next row) under column (1) as input; the output as rice is entered in column (2) of the same row using appropriate conversion factors. This would be the procedure for derived products.

Changes in stocks (column (3))

Changes in stocks relate to the reference period at all levels between production and retail levels, i.e. changes in Government stocks, in stocks with wholesale and retail dealers, importers, exporters, manufacturers, transport and storage enterprises and on farms. In practice, however, data available with most developing countries relate only to stocks held by Governments.

+ relates to net increases in stocks.

− relates to decreases.

Gross imports (column (4))

This item covers all commodities imported into the country, as well as commodities derived therefrom and not separately included in the food balance sheets. Countries generally depend on customs returns for determining quantities imported — these, however, are restricted for those quantities liable for payment of customs duty and therefore exclude food aids and gifts. Care should therefore be taken to include all commodities imported whether liable to customs duty or not.

Weight for imports means net weight, i.e. excluding the weight of containers.

Supply (column (5))

Supply is given as Production + Gross Imports + Decrease in stocks, or Production + Gross Imports − Increase in stocks, depending whether there has been a decrease or increase in stocks during the period.

FOOD BALANCE SHEET

Country: .... Population (mid-year): 121 632 (thousand) Year: 1972
In 1000 metric tons unless otherwise specified

COMMODITY	Production						Domestic Utilization							Per Caput Consumption				
										Manufacture for								
	Input	Output	Changes in stocks	Gross imports	Supply	Gross exports	Total	Feed	Seed	Food	Industrial use	Waste	Food	Kg per year	Grams per day	Calories per day number	Proteins per day grams	Fat per day grams
	(1)	(2)	(3)	(4)	(5)	(6)	(7)	(8)	(9)	(10)	(11)	(12)	(13)	(14)	(15)	(16)	(17)	(18)
CEREALS																		
Paddy		18131			18131		18131	181	319			725	(16906)	–	–	–	–	–
Paddy/Rice	16906	11496	–363	748	12607		12607					252	12355	101.58	278.30	1002	18.7	2.0
STARCHY ROOTS AND TUBER																		
Potatoes		124			124	3	121		12			6	103	0.85	2.33	2	–	–
PULSES, NUTS AND OIL SEEDS																		
Groundnuts (in shell)		470			470	1	469		36	30		23	(380)	–	–	–	–	–
Groundnuts (shelled)	380	228			228	13	215						215	1.78	4.88	27	1.2	2.1

FOOD BALANCE SHEET

Country: .... Population (mid-year): 121 632 (thousand) Year: 1972
In 1000 m/t unless otherwise stated

COMMODITY	Production						Domestic Utilization							Per Caput Consumption				
										Manufacture for								
	Input	Output	Changes in stocks	Gross imports	Supply	Gross exports	Total	Feed	Seed	Food	Industrial use	Waste	Food	Kg per year	Grams per day	Calories per day number	Proteins per day grams	Fat per day grams
	(1)	(2)	(3)	(4)	(5)	(6)	(7)	(8)	(9)	(10)	(11)	(12)	(13)	(14)	(15)	(16)	(17)	(18)
MEAT																		
Cattle/meat		152			152	20	132						132	1.09	2.97	7	0.5	0.5
Cattle/offals		23			23	3	20						20	0.16	0.45	1	0.1	0.1
FATS AND OILS																		
Groundnuts/oil (in shell)	30	9			9		9						9	0.07	0.20	2	–	0.2
Cattle/fats	132	4			4		4						4	0.03	0.09	1	–	0.1
													Total			1984	42.2	28.8
													Vegetal			1929	37.2	26.2
													Animal			55	5.0	2.6

Supply/Utilization accounts (Food Balance Sheets)

Gross exports (column (6))

This item covers all movements of the commodity in question out of the country during the period. Weight means net weight.

Domestic utilization (column (7))

The quantity here is the difference between supply (column (5)) and gross exports (column (6)). This is the quantity available for domestic utilization. This item includes utilization for feed, seed, manufacture and food, and also quantities wasted. Thus, column (7) = sum of columns (8), (9), (10), (11), (12) and (13).

Domestic utilization is sometimes described as available supply.

Feed (column (8))

This is the amount fed to livestock during the reference period, whether domestically produced or imported.

Seed (column (9))

This item comprises all amounts of the commodity in question used during the reference period for reproductive purposes such as seed, cane planted, eggs for hatching and fish for bait, whether domestically produced or imported.

Manufacture (columns (10) and (11))

This item refers to processing of foodstuffs for food and industrial uses. Examples for food are oilseeds into edible oil and fresh fish to salted, dried and canned fish. In this case, the processed foods reappear in the next row in the food balance sheet.

An example for industrial use is, oil for soap. In this case, the processed items will not appear again in the food balance sheets.

Waste (column (12))

This comprises amounts of the commodity in question and of the commodities derived therefrom not further pursued in the food balance sheets, lost through wastage during the reference period at all stages between the level at which production is recorded and the retail level, i.e. wastage in processing, storage and transportation.

Food (column (13))

Column (13) = columns (7) − (8) − (9) − (10) − (11) − (12).

If the item of food is such that it is consumed directly like potatoes and yams then this column is entered. Otherwise (i.e. when consumed after processing) column (13) is not entered in the row of the primary product but entered in the next row under column (1) input and converted into the processed product under column (2) by suitable conversion factors and after dealing with columns (3) to (12), entered under column (13) in this row.

Per caput consumption (columns (14), (15), (16), (17) and (18))

The five columns concerned give estimates of *per caput* food supplies available for human consumption during the reference period in terms of quantity, calorie value and protein and fat content. *Per caput* consumption is given in kilograms per year and grams per day. Calorie supplies (column (16)) are reported in kilocalories (Calories) per day and protein and fat consumption in grams per day.

Column (14) is obtained by dividing column (13) by the de facto population at the mid-point of the reference period. Column (15) is obtained from column (14). Columns (16), (17) and (18) are obtained from column (15) by using conversion rates of the country concerned. FAO's international food composition tables could be used after suitable modifications for the country concerned.

9.5 Guidance on Filling the Balance Sheets

Primary products which are consumed directly like potatoes have only one row in the sheets. Derived products will have more than one row. For example, for paddy there will be two rows — one for paddy and the other for paddy/rice. For the first row, columns (2) to (12) are only filled. In the second row, all the columns (1) to (18) have to be filled. For some products like groundnuts there are three rows. In the first row, columns (13) to (18) are not filled as for paddy. In the second row, what is consumed directly as shelled groundnuts is dealt with. There is a third row for groundnuts converted as oil, not directly below the second row but later under "Fats and Oils" where column (1) indicates the quantity given under column (10) in the first row for manufacture. Balance columns in this row are filled for utilization as oil (see example in section 9.6). For meat there will be three rows — first row for meat, second row (not necessarily below the row for meat) for offals, and the third row under

Supply/Utilization accounts (Food Balance Sheets) 81

"Oils and Fats" for fat.

No dependable data are available in developing countries for feed. Livestock officers may indicate the quantity of feed as percentage of production. Sample surveys should be made to arrive at dependable figures.

Seed rates are available in all countries — in the case of crops as kilograms per unit of land. As different seed rates are used for the same crop even within countries, it is useful to work out an average for the whole country for every crop. The quality of data used can be improved by sampling surveys — some countries make use of crop cutting surveys for collecting data on feed, seed and waste on a sub-sample basis.

Data for manufacture are collected by Governments from manufacturing organizations but these do not always include data from small organizations. As such these data need modifications for use in the balance sheets.

Waste is generally given as a percentage of production. For dependable data sample surveys especially by marketing organizations are needed.

For processed commodities, dependable extraction rates should be determined for converting from the primary products. These rates are best determined in cooperation with the authorities of the processing plants (e.g. rice mills, flour mills, tea factories, sugar factories). The rates have to be revised periodically with changes in the varieties of the primary products and the processing methods.

9.6 Example

An example is given here of the food balance sheet of a developing country for 1972. Some food items have been selected and the forms filled according to available data. The mid-year population of the country was 121,632,000. The details of the working are given below for the selected items.

Paddy

The harvested production was 18,131,000 metric tons (m.t.). From this figure deductions of 181,000 m.t. for feed (taken as 1% of production),

82 Agricultural Statistics: A Handbook for Developing Countries

319,000 m.t. for seed (at the rate of 40 kilograms (kg) per hectare of planted area)[1] and 725,000 m.t. for waste (taken as 4% of production) are made leaving a balance of 16,906,000 m.t. This figure is now entered in the next row column (1) for conversion to rice. Taking the extraction rate as 68% from paddy to rice, column (2) becomes 11,496,000 m.t. Adjusting for change in stocks, imports and waste in the rice form (2%), column (13) gives the rice available for human consumption. Column (14) is obtained by dividing column (13) expressed in kg by 121,632,000 (mid-year population). Column (15) is obtained by expressing column (14) in grams and dividing by the number of days of the year (in this case 366 – being a leap year). Columns (16), (17) and (18) are the product of column (15) and the appropriate conversion factors to Calories,[2] proteins and fat respectively. (For rice they are 3.6, 6.7% and 0.7%.)

Potatoes

For potatoes, the seed rate used is 700 kg per hectare, waste 5% and conversion rates of 0.7, 1.7% and 0.1% for calories, proteins and fat.

Groundnuts

The production is generally given with the shell. In the first row, deductions are made for seed, manufacture for food (as oil which will be entered later under fats and oil) and waste. The balance is repeated in the next row column (1). The extraction rate of groundnuts in shell to shelled groundnuts is 60% and thus the figure in column (2) is 228,000 m.t. After allowing for export, column (13) is entered and other columns are arrived at using suitable conversion factors.

Meat

Cattle meat is the item given – the weight in column (2) is the dressed carcass weight. The weight of offals is given as a percentage of carcass

---

[1] Data on planted areas are always available.

[2] Calories here mean kilocalories.

weight — in this case 15%. The fats from cattle will be entered later under fats and oils.

Fats and Oils

The first item given here is groundnut oil. Column (1) has the figure given under column (10) of groundnuts (in shell). Column (2) is obtained from column (1) by multiplying by 0.32 (32% is the extraction rate of oil from groundnuts in shell).

The second item given is cattle fat. Column (1) gives the carcass weight given under column (13) of cattle meat. Column (2) is obtained from column (1) by multiplying by 0.03 (3% is the slaughter fat extraction rate of cattle meat). The total calorie, protein and fat intake are given at the end of the sheets with breakdown in vegetal and animal origin.

CHAPTER 10

Census of Agriculture

10.1 General Remarks

Historically a census meant a census of population, counting all individuals in a country at a given point of time with important particulars on sex, age and occupation, unit of enumeration being a human being. The concept was later extended to housing with enumeration unit as a house, industry with enumeration unit as an industry and agriculture with enumeration unit as an operational agricultural holding[1] (with items such as area, farm population, livestock, machinery, implements) at a point of time. In other words a census of agriculture meant taking an inventory of agriculture for the whole country at a point of time. This concept gradually expanded beyond that of an inventory; and point of time extended to an agricultural year to cover items like area under crops, yields and employment.

Today the concept of an agricultural census has expanded:
    (a) To provide an integrated picture of a country's agricultural structure;
    (b) With time reference extended beyond a point of time;
and complete enumeration modified to part complete enumeration and part sampling.

With the experience gathered in the several countries which participated in the three decennial agricultural censuses of 1950, 1960 and 1970, a methodology for a census of agriculture can be evolved under the following heads:
    I. Objectives;
    II. Scope and coverage;
    III. Concepts and definitions;
    IV. Planning and procedure (including census design);

---

[1] What is meant as a holding is described later in the chapter.

V. Analysis, tabulation and final reports.
It is convenient to deal with the subject under these five heads.

10.2 Objectives

The objectives of an agricultural census can be broadly stated as providing:
 (a) Data for clarifying the social and economic factors of a country's agricultural structure by inter-relating various characteristics of the holdings, e.g. size of holding and type of holding on the one hand, and factors such as fragmentation, tenure, land utilization, crop patterns, use of fertilizers, number of different kinds of implements and machinery, farm population and labour force on the other.
 (b) Detailed agricultural data,[1] such as number of holdings, total area under holdings, basic pattern of land utilization, area under crops and extent of irrigation at the level of the smallest administrative division of the country.
 (c) A bench mark for improving reliability of current agricultural statistics and for assessing future agricultural development.
 (d) A technical and organizational apparatus for a sound permanent system of agricultural statistics by utilizing the census data as a frame for projecting agricultural surveys and strengthening the agricultural statistics offices with the census experience and trained personnel.

The first may be considered the major objective of an agricultural census. The emphasis is at present on studying the various characteristics of farms and farmers, income distribution among different classes of farmers and other allied topics for the entire country and its sub-divisions for socio-economic planning and implementation of projects for increased agricultural production. For such purposes the census is the only source of data by small regions. From this point of view, quantitative data such as area and yields may be regarded as of secondary importance in a census. The collection of this information is, however, essential for economic

---

[1] Concept of "inventory" of agriculture.

classification of farms. (The data secured by interviewing the farmers in a census are of doubtful accuracy. There are other methods such as annual surveys to improve on the dependability of such quantitative data which should supplement census data.)

The census provides a bench mark against which future agricultural development is to be measured. It also provides a periodic inventory of national agricultural resources and their geographic distribution.

The census, by itself, accomplishes a limited but important task of providing a broad picture of the socio-economic structure of a country's agriculture; it can also supply a certain amount of data for the country's agricultural statistics. The decennial census, though being no substitute for securing current agricultural statistics, offers an excellent base and framework for future surveys on needed agricultural statistics and for technical and organizational training of the personnel required for the surveys and for operating the country's system for collection of current agricultural statistics. Unfortunately not many countries have worked it this way.

10.3 Scope and Coverage

10.3.1 *Scope*

The programme recommended by FAO to the countries includes collection of data on the following items:

(1) Number of agricultural holdings and their principal characteristics, such as size, form of tenure, utilization of land by broad classes, type of holding (whether producing mainly for home consumption or for sale);

(2) Area under crops and volume of production of the principal crops;

(3) Number of livestock and volume of production of some livestock products;

(4) Employment in agriculture;

(5) Farm population;

(6) Agricultural power and machinery and general transport facilities;

(7) Irrigation and drainage;
(8) Use of fertilizers and soil dressings;
(9) Wood and fishery products obtained from agricultural holdings;
(10) Extent to which agriculture is associated with other industries.

The scope of the agricultural census for a country would depend primarily on the country's need on agricultural information and conditions governing its collection. There is a general tendency among countries to include a large number of items like those indicated above. It would, however, be wise to limit the scope to those items in consultation with the users, for which the needs for data are greatest and at the same time the items lend themselves for collection of reasonable information as part of the census. Further, whatever be the needs for data, only those data are worth collecting under given conditions, where there would be a prospect of securing information of reasonable accuracy.

10.3.2 Coverage

Ideally the census should cover all agricultural holdings, irrespective of size, economic status or location. Practical considerations make it necessary to limit the coverage omitting such portions (a) where the cost or difficulty of enumeration is unduly disproportionate to the contribution of these portions to the country's agriculture and (b) where their inclusion is liable to show a distorted picture of the agricultural structure. In the first category would be some urban areas, nomadic tribal areas, inaccessible areas and in the second category would be very small holdings and kitchen gardens.

(a) For *geographical coverage*, highly developed urban areas can be omitted for collecting data on crops while these areas cannot be omitted for livestock purposes (poultry farming, dairying). For example, in the 1962 Census of Agriculture in Sri Lanka urban areas under the big municipalities as well as some portions of urban council areas where there was little crop agriculture, were omitted for collecting data on crops. However, for the livestock count, no areas were omitted.

Cultivation of nomadic people can also be omitted for purposes of the census in view of the many difficulties concerned like inaccessibility and excessive costs and also due to the fact that this cultivation accounts for an insignificant part of the country's agriculture. Nomadic livestock cannot be neglected as their contribution to the livestock of the country concerned is considerable.

(b) Should there be a *minimum size* for holdings to be included in the census or not, needs careful consideration. There is no question of omitting small holdings as they predominate in numbers although not accounting for a major portion of a country's agriculture. If, however, no minimum limit is placed, house compounds with a few fruit trees, kitchen gardens growing some vegetables, and households keeping a cow or two or a few birds for eggs would all become eligible for inclusion. This would increase the cost of the census and seriously distort the size distribution of holdings, thereby creating a wrong picture of a country's agricultural structure. Thus countries generally fix a minimum size, e.g.:

in Indonesia   1/10 hectare for crops
                or 10 cattle/buffaloes/horses   } for
                or 50 goats/sheep/pigs   livestock
                or 100 poultry
in Sri Lanka   1/8 acre

10.4 Concepts and Definitions

A clear understanding of the basic concepts concerning the various aspects of the census is important for selecting units for enumeration and for eliciting information on the different items included in the census. In defining concepts, it is necessary to visualize the possible uses of the census results and the consequences of the definitions adopted on the type of results provided by the census.

Once the items for the census are settled, they will have to be precisely defined, giving no room for ambiguity. The definitions should be such that the results are comparable over time and space. If, however, a change in an earlier concept or definition is considered essential, a link with the earlier definition has to be provided. For census results to be comparable over space, concepts and definitions should not only be uniform within a country but also, as far as possible, with countries in the region. Further, there should be consistency between the various items in concept. For example, in a land holding one should be clear of what is meant by a parcel, or a field or a plot.

The operational agricultural holding is the key item, and basis for enumeration in an agricultural census. It is defined as all the land which is used wholly or partly for agricultural production and is operated as one technical unit by one person alone, or with others, without regard to title, legal form, size or location. Establishments and other units not including any agricultural land but producing livestock or livestock products are also considered as holdings (whether they are located in rural or urban areas). A technical unit is the unit which, under the same management, has the same means of production, such as labour force, machinery and animals.

For definition of other items, references should be made to FAO publications on World Census of Agriculture.

As regards *time reference*, information collected for the census refers to one year or to one week or to a specific date according to the subject investigated. The duration of the enumeration should be rather short and generally should not extend beyond one month, in order to avoid omissions or duplications because of changes in figures on some items collected in the census caused by variability of some events, e.g. changes in employment in agriculture, movement of livestock. In cases of enumeration with more than one round, the observations made, relate to each round separately.

An indication of time reference, by some items is given below:
(1) A specific year, such as the last agricultural year or the year preceding the date of enumeration:
 items: aggregate area under crops;
 crop production;
 production of livestock products;
 use of agricultural power on the holding;
 agricultural machines used on the holding;
 general transport facilities;
 area of land irrigated;
 fertilizer and soil dressings.
(2) The *week* preceding the date of enumeration item:
 employment in agriculture.

(3) A *specific date*: usually considered the day of enumeration or some other day close to this day.
items: holding;
holder;
tenure of holding;
farm population;
agricultural machines owned by the holder.

10.5 Planning and Procedure

Census design. The census is conceived in principle as the collection of data for all individual agricultural holdings by direct enumeration. However, some countries have conducted sample censuses as a substitute for complete enumeration, that is, while the coverage was the entire country, only a sample of agricultural holdings was enumerated. The sample census has several advantages. It requires a smaller number of field staff and consequently better trained and better supervised. The cost of the census is lessened and quality of data collected improved. Further, the burden on the agricultural population is less as only a fraction of them would be called upon for information. There will be a saving on time in processing, tabulation and presenting the final report. Thus the sample census has its attractions.

On the other hand a sample census has its limitations. It cannot provide data by small administrative units which are required for agricultural development and planning. Further, sampling cannot provide accurate data for detailed cross tabulation on a district basis and the numbers available for individual cells in a deeper cross-classification tend to become too small and consequently subject to large sampling errors. When a census is taken by complete enumeration, it would serve a very useful basis and framework for various surveys (for estimation of crop areas and yields, livestock products, cost of production, etc.). This advantage is lost if the census is a sample census. In view of these reasons, it is sometimes felt that complete enumeration is a desirable undertaking even if the results are likely to be of imperfect quality and have a limited scope.

Against this background, a decision has to be taken for complete enumeration or sample census or a combination of the two. Some countries completely enumerate large holdings and take a sample census for small holdings. In Taiwan, for instance, complete enumeration was attempted for main crops and sampling for minor crops. A combination of the two methods has been found to meet the requirements of many countries in their decennial censuses and should therefore meet the requirements for the future.

Census planning. It is convenient to consider planning as consisting of three phases. Once the design of the census and data to be collected are settled, a suitable questionnaire has to be drafted. At the same time the tabulation plans should be prepared in consultation with the data processing unit as this helps to define and delimit the content of the questionnaire clearly.

In the first phase, this questionnaire should be tested in the field in as many localities as possible representing the various agro-economic and social conditions of the country. This test should help to find out the effectiveness of the questionnaire in eliciting the desired information and of the procedures followed in the field as the test involves interviews with typical respondents who would be normally interviewed at the census. On the basis of this test, the questionnaire may be finalized.

In the second phase, a pilot census will be carried out in a few selected areas representing different agricultural conditions of the country. Each area should as far as possible be of the size of the smallest region for which results of the actual census are to be tabulated. The pilot census is thus really a full dress rehearsal of all the processes of the actual census. It will test the adaptability of the various forms, indicate time and personnel required for the several operations. The various tables should be prepared and sampling errors calculated. Based on the experience in the pilot census, census design, forms and questionnaires, staff and budget requirements could be suitably modified to meet the needs of the actual census.

The third and final phase would involve all operations of the actual census such as training of field staff, publicity, printing of all forms and questionnaires, field enumeration, tabulation and publication of the results.

Procedure. In order to prepare a frame for census enumeration, systematic pre-listing of households for the whole country is necessary. For this purpose, the whole country has to be divided into small areas. Such a small area can be a village but where villages are large, each village can be sub-divided into two or more enumeration blocks (EB). Where the population census has preceded the agricultural census, lists of enumeration blocks are readily available with well defined maps showing EB boundaries. These enumeration blocks can then be distributed to the enumerators.[1] Where a population census has not preceded the agricultural census, it is worth the while to demarcate the villages into enumeration blocks.

In each EB the enumerator will cover all households and mark those eligible as agricultural households. If a population census has preceded the agricultural census, lists of households would be readily available and with foresight at that census if a question had been asked whether the household is an agricultural one or not, lists of agricultural households could be picked out and the enumerator's task is to bring such lists up-to-date.

When the pre-listing operation is completed, households for actual enumeration are made ready. The actual enumeration may be a complete one or a sample or part complete and part sample. The sample households are selected at this stage and each enumerator allotted his households for interview.

With the experience gathered in the pilot census and the training received by the enumerator and with the questionnaire finalized after testing in the field with farm households, the enumerator now has to face the respondents (holders or their representatives) with the questionnaires. Though the questionnaire has been so prepared as to enable the enumerator to secure precise and unambiguous information from the respondent on various items, it may not be possible to do so mechanically in all cases. Sometimes he will have to discuss each item with the respondent like a conversation. In such circumstances it is a good practice to provide the enumerator with a field notebook to make his notes at the spot and later fill up the questionnaire. (Instead of a notebook a few blank sheets at the end of each questionnaire would serve the purpose.) For example, it is not always possible for a holder to give a direct answer for the area of an operational holding. In such cases the enumerator has to find out all the land with which he is connected, deduct all land which he may

---

[1] Generally the number of households for an EB is so chosen that one enumerator can handle one EB.

own but does not use himself and add all land he does not own but uses.

Cases have come up in past censuses where respondents failed to give information on lands operated outside the village. The enumerators should make a special request regarding lands outside the village. Likewise special requests should be made for minor enterprises of small holders like growing a few fruit trees and vegetables or rearing a few goats, pigs and poultry.

Supervision of enumeration work is an essential part of the census. Supervisory staff should go round during the enumeration and do some test checks in the field. When the completed questionnaires are received by the supervisors they should check the forms before dispatching to headquarters. If any forms have to be returned to the field, it is much easier to return them at this stage from the supervisors' offices than later from the head office.

10.6 Analysis, Tabulation and Final Report

Before analysis and tabulation commence, a critical examination of the questionnaires received from the field is essential. This is done after arranging them in suitable batches and editing in order to fill in omissions and correct errors by referring back to the field when necessary.

A detailed list of tables, their type and format would have been finalized in consultation with the data processing unit or office during the pilot census stage. Hence at the present stage the questionnaires after scrutiny by the census unit should be sent to the data processing unit for analysis and tabulation.

In the tabulation plans, the FAO has recommended to countries that tabulation of certain important items by size classes as compulsory and the balance items as optional. For example, in the section for holding, holder, tenure and type of holding, FAO recommends among other compulsory tables, a table for number, area, mode of operation and fragmentation of holdings by size, leaving it optional for a table like fragmentation of holdings, by tenure and by size of holding.

The tables, decided upon by the countries would be prepared by the data processing unit and made available to the census unit.

The census unit should endeavour to publish a preliminary report within six months of the completion of the census embodying important preliminary results such as areas and production of principal crops by provinces/districts and other information needed for economic planning. The final report should be published as early as possible but not later than two years. Besides, special reports can be prepared on items of importance to the country in the meantime.

The use of the census results depends upon the degree of accuracy and therefore it is briefly described here. When a census is based on a sample, the results are subject to sampling errors. Fortunately these errors can be controlled and can be reduced to a desired level. Provided the sampling errors are small, the usability of the numerical results is not affected to any significant extent. There are, however, other errors (called non-sampling errors) which affect all census results whether by complete enumeration or by sampling. Further, unlike sampling errors which affect the results symmetrically, non-sampling errors are mainly in one direction and there is no way of determining their magnitude. Hence in all census operations, great care should be taken to reduce such errors to a minimum. In the census operation the interviewer and the respondent both contribute to the error in the results. The interviewer may be poorly informed of census objectives, technically ill-equipped, inadequately trained and not sympathetic to the respondent. These weaknesses can be remedied or controlled with a good questionnaire, proper selection and training of the enumerator, with adequate facilities and incentives and above all exercising close supervision of his work. The respondent may be uneducated or illiterate lacking in quantitative understanding, and deeply suspicious of the enquiry. These may be partly overcome by proper educational publicity and assurance from people whom respondents generally trust that the information collected would be treated as confidential and not used against them particularly for tax purposes.

However skilful the interviewer may be there is a certain amount of under-reporting particularly for area of the holding, acreage under crops, number of livestock, etc. The census results are consequently biased.

Thus an evaluation of the census results for the information of the users as well as for guidance for future censuses is desirable. For this evaluation, use of external check data, consistency over time, internal consistency among the various items of the census are some of the possible approaches. As an aid for checking internal consistency, the importance of a village questionnaire may be emphasized. Countries are considering a post-enumeration survey as a method of evaluating census results.

The post enumeration check is done by means of small sample surveys carried out soon after the census, utilizing better personnel and a restricted questionnaire with only some key items. The object of this survey is to check on the completeness of coverage and accuracy of the data reported during the census. Every item in this survey will have more detailed questions than for the census, e.g. instead of area as a whole, parcelwise or plotwise areas will be asked for. There has been some doubt on the efficacy of such surveys if the data are collected by interview alone. As such these surveys should include use of physical methods for measurement of area and yield and an actual count of livestock.

There is a school of thought suggesting the use of these methods in conjunction with the census while the census is in progress. There will be savings in expenditure and the supervisory staff could be utilized and part of their work could be to make physical measurements on a sample of farms. There are two definite advantages in this method. Firstly the work of the enumerators would be improved. Secondly the prospects are better for utilizing the check data for adjustment of census results than for a post enumeration survey as the data collected are too scanty for such adjustment.

It is up to the country to decide which type of check should be done for an evaluation of the census results.

CHAPTER 11

Survey and Sampling Methods

11.1 Scope

Though surveys and sampling have been mentioned in the book on several occasions in earlier chapters, it is considered worthwhile to put in one place their essential features. Some repetition seems unavoidable. It was mentioned in the previous chapter that the word census was confined to a complete census till it was modified later to include a sample census. In a similar way the word survey was confined to a sample survey which, however, is a mistake as it embraces complete survey as well.

11.2 Surveys

For an investigation involving the collection of elaborate information, the term survey is generally used. Surveys are conducted involving complete enumeration and/or sampling. Normally one does not resort to complete enumeration in all its aspects. The general practice is for sampling alone or for a combination of part enumeration and part sampling. While some surveys are conducted regularly to provide agricultural data on an annual or seasonal basis, others are conducted as and when data are needed.

In surveys both enquiry and objective methods are used. Though it is preferable to use objective methods as stressed in earlier chapters, there are limitations. These may be due to lack of sufficient funds and trained personnel. Further there are some investigations which do not lend themselves to objective measurements. For example, in cost of production surveys, it is not possible to use objective methods for all items. In food consumption surveys, objective methods are not possible for some items unless householders cooperate to weigh some of the foodstuffs before cooking. In price surveys, objective methods are not possible unless some of the products are actually bought.

Due to these reasons, a combination of enquiry and objective methods should meet most of the needs of a survey. Sometimes objective methods are restricted to sub-samples, while enquiries are conducted for the whole sample.

As for census there should be a methodology for surveys. The methodology for census described in Chapter 10 applies for surveys with suitable modifications and is therefore not repeated here. The main headings, however, are repeated with some modifications:

I. Objectives;
II. Scope and coverage;
III. Concepts and definitions;
IV. Planning and execution (including design);
V. Analysis, tabulation and final report.

These headings should give some guidance in the conduct of surveys.

11.3 Sampling Methods

11.3.1 *General*

Sampling methods are used in agricultural statistics for the same reasons as in other fields of statistics. Namely, by reducing the volume of the material to be dealt with, (i) the cost of the work is reduced, and (ii) makes it possible to improve the quality of data because of the reduced number and better selection of personnel.

In addition to these general reasons, there are some special features of agricultural statistics which make the use of sampling convenient and sometimes unavoidable. One of these features is the extremely broad programme of agricultural statistics, consisting of areas of many crops, their yield, livestock statistics with all its distribution of age, sex, and race, and statistics of livestock products, horticulture, agricultural machinery, labour force, etc. As many characteristics in each of the field listed require separate surveys, the most efficient way to deal with such a broad programme is to have recourse to sampling consequently reducing the pressure on the limited resources of developing countries.

Another feature in agricultural statistics, which has already been mentioned in the book, is that farmers are frequently not in a position (due to ignorance, fear of taxation or lack of appreciation of quantitative information) to supply the information requested in surveys. While demographic surveys in most cases ask for simple facts which would be remembered such as the number of persons in a household, sex and number of children born, agricultural characteristics are either forgotten fast or never precisely known such as production of crops, quantities of fertilizers used for various crops, production of dairy products during a specific period, number of days worked off the farms. A vague memory and knowledge of such characteristics on the part of the farmers make it necessary in some cases to use methods and techniques which are independent of the farmers themselves. Such techniques might be the system of special book-keeping, direct observations by specially selected personnel for that purpose or physical measurements. In view of the cost of the application of these techniques and the specialized knowledge needed, sampling is again the most useful technique.

Further, units of agricultural surveys would be spread over a very large territory. For example, fields under some crops would be scattered over the whole country (e.g. paddy fields in Malaysia, Indonesia and Sri Lanka). If, in the survey, the intention is to visit these fields (like crop cutting), there is often no other way than the use of sampling as a means to reduce travel by the personnel needed and the time necessary to carry out such surveys.

In developed countries, sampling primarily opens the way to savings and better accuracy while other techniques may be equally applicable. In developing countries, however, the use of sampling methods is sometimes the only way to collect agricultural statistics.

11.3.2 *Application*

There is widespread use of sample surveys in agricultural statistics. Sampling is so important that in Agricultural Statistics Divisions of some developing countries, there are sampling units to handle such work. Some of the applications of sampling to development of agricultural statistics

Survey and Sampling Methods

have been given in earlier chapters.

Some of the important fields in agricultural statistics where sampling is used, are given below:
(1) Areas;
(2) Yields;
(3) Fruit production;
(4) Livestock;
(5) Livestock products;
(6) Food consumption surveys;
(7) Checking quality of statistical data;
(8) Agricultural experimentation.

11.3.3 *Frame, stratification and sample design*

11.3.3.1 *Frame*

All sampling demands a sub-division of the material to be sampled into units, called *sampling units*, which form the basis of the actual sampling procedure. These units may be:
(a) Natural units of the material
 e.g. individual in a human population
  individual animals in livestock;
(b) Natural aggregate of such units
 e.g. households, agricultural households, agricultural holdings;
(c) Artificial units
 e.g. crop cutting plots.

It is not always necessary to make an actual sub-division of the whole of the material before selection of the sample, provided the selected units can be clearly and unambiguously defined. Thus, with sampling units with crop cutting plots, there is no need to demarcate all these plots; they can be defined by coordinates, and the selected areas demarcated after selection.

Clear and unambiguous definition demands the existence or construction of some form of *frame*. In the use of sampling for livestock numbers, for instance, with livestock holdings as sampling units, there must be available a list of all livestock holdings, and this list must be such that any livestock holding selected from it can be unambiguously located. In area sampling from maps, the maps must be such that the selected areas can be unambiguously defined on the ground.

Any aggregate of values is termed a *population*, and consequently the whole aggregate of sampling units into which the materials is divided is known as the population of sampling units. If the sampling units are aggregate of the natural units of the material, these natural units will form a further population which must be distinguished from the population of sampling units.

In multi-stage sampling there is also a hierarchy of sampling units, primary or first-stage, second-stage, etc. corresponding to the different sampling stages, and each set of units will form its own population of units.

Sampling units may be of the same or differing size. They may contain the same or different number of natural units. The entire procedure of sampling is simplest when the sampling units are of approximately the same size and contain about the same number of natural units. However, variation in size of the sampling units or in the number of natural units they contain is inevitable if the natural units themselves are of widely differing size.

In the construction of a frame, there are two approaches:
(1)   As made up of human or business units;
(2)   As consisting of areas of land.

Both approaches have their advantages and disadvantages. The first category can be one of farmers (growing the crop/rearing livestock) or of dairy or processing plants. Any form of enquiry would need the cooperation of the farmers or those who control the plants as the case may be. Hence the first approach has its advantages. If, however, the cooperation of the farmer is not required, area frames are quite suitable. A serious disadvantage of the area frame is that there is no simple and reliable means of knowing without visiting all the areas whether they include agricultural activities or not.

A good frame of agricultural households can be prepared immediately after a population or agricultural census. (If a population census alone is taken, a question can be inserted in the questionnaire to find out whether a householder is an agricultural householder or not.) These lists can be brought up-to-date every year and would be available at village level. Alternatively where tax is levied for land or for irrigation, lists of farmers using the land or water are prepared and a frame can be constructed this way.

Where cultivation is on a highly organized basis, lists of fields would be available and an area frame can be constructed.

A serviceable frame can be constructed in any one of the ways mentioned above depending upon the circumstances of the country and the objectives of the survey, keeping in mind that a frame suitable for one characteristic of the material to be sampled need not necessarily be suitable for another characteristic of the material.

11.3.3.2 *Stratification*

Stratification has been described at length in section 4.2.2 and therefore need not be repeated here. Generally in agricultural surveys, stratification is by administrative, geographical, or agricultural regions, and sub-stratification by sub-administrative regions, by size of villages or by size of processing plants (dairy plants, tea factories, etc.) or size of collecting centres depending on the subject of the survey and the system of collection of data.

11.3.3.3 *Sample design*

Some sample designs have already been described. These are for determining yields, areas, livestock numbers, and livestock products. They are all stratified multi-stage sampling. In all these cases stratification has been by administrative or geographical or agricultural regions and most surveys will continue to use these stratifications.

The primary sampling units in most cases are either villages or communes or artificially created clusters (say of a certain acreage under a crop where area or yield rate has to be determined).

The second stage units for crops generally are parcels or fields. Lists of parcels/fields are prepared with the assistance of farmers in most cases while in other cases lists can be prepared directly from maps which are available with the cultivations marked.

For surveys on livestock and livestock products, the second stage units are livestock holdings or farmers.

The third stage units for crop yields would be the crop cutting plots while for cropped areas generally there are no third stage units. The latter is also the case for surveys on livestock and livestock products. In these cases the sampling stops at the second stage.

Sampling becomes simple when dealing with big organizations. For example, where milk collection is highly organized, the frame consists of collecting centres. Here the sample units are stratified by size of the collecting centres (i.e. by volume of milk collected) and sampling is a simple random one for each stratum.

In most of the cases mentioned, for distributing the sampling units within a stratum, a constant proportion has been used. There are, however, cases where a variable proportion gives greater precision than a constant one for the same number of sample units in the stratum. An important application is for material stratified into size-groups where considerable gains in precision will result if different sampling fractions are used for the different size-groups. The greatest precision for a given number of units will be attained if the sampling fractions used are proportional to the standard deviations of the size-groups. But information regarding standard deviations of the size-groups is not always readily available. In such cases, it has been found that for many types of material, a good approximation would be to take the sampling fractions proportional to the mean sizes of the size-groups. When variable sampling fractions are used, it will be necessary to weight the contributions of the different size-groups in the correct proportions to get the stratum estimates.

An example of the use of different sampling fractions can be seen in section 4.4. Here for the second stage of sampling, sampling fractions of 1/10 and 1/25 were used for small holdings and gardens respectively, small holdings being less than 20 acres but more than 1 acre and gardens 1 acre or less in extent.

11.3.3.4 Drawing samples

When the number of sampling units for the survey is settled, these are allotted to the several domains. In each domain the units allotted should be distributed through the various stages in accordance with the design. The actual drawing of the samples should be done by officers, other than enumerators and never by the enumerators who attend to the field work (filling the questionnaires, doing the physical measurements, etc.) Experience has shown that some enumerators for their convenience select easily accessible units and do not follow the instructions regarding selection. Therefore it is preferable for senior officers to draw the sample, preferably the officers in charge of the domains. The selected units should then be distributed to the various enumerators by the supervisors.

Supervision of enumerators' field work is essential to improve on the quality of data collected. This need not necessarily be by the supervisors alone. Any senior technical officer who is about the place of enumeration could help. For instance, in Sri Lanka when a crop cutting is finalized by the enumerator in consultation with the farmer, a telegram is sent at least two days earlier to Colombo (the capital) of the venue, date and time of the crop cut. The Statistics Department in Colombo would then instruct any senior officer who would be about the locality to visit the field and witness the conduct of the crop cut. This type of arrangement has a salutary effect on the enumerators to conduct their work well as any senior officer might witness the crop out.

On completion of field work, the data collected would be sent to the supervisors who would scrutinize the data and if necessary visit some localities for purposes of checking any doubtful data before despatching to the Head Office. After scrutiny in the Head Office, the data are analyzed, and tabulated, and finally a survey report is prepared and made available to the authorities and the public.

11.3.4 Bias

As bias can arise from the procedure of selecting the sample, it is briefly mentioned here. A procedure can be biased either because it was conceived and formulated as such or it became biased in its implementation. The effect of a selection bias can be removed from the estimates by computational adjustment of sample data. If the selection is biased in that some classes or groups of the population are under-represented while some others are over-represented, sample data are weighted by the respective known class or group frequencies from an earlier survey or census, in the computational adjustment.

There is another method for removing the effect of the bias, which is by the adjustment of the structure of the sample. This is carried out by changing at random the number of units in the different classes or groups of the sample so as to reach the same frequency distribution as the one in an earlier survey or census.

CHAPTER 12

Price Statistics

12.1 Importance

Statistics of prices connected with the agricultural industry are very important as price relations play a fundamental role in the formulation of agricultural development plans and related decisions of an economic nature. These relations largely determine the type and volume of the productive activity in agriculture. However, many developing countries have not accorded the compilation of meaningful and adequate price statistics the same attention as that given to other statistics such as cropped areas and yield rates.

Price statistics should consist of a systematic collection, processing and publication of price data with the purpose of using them for:
  (a) Description of those aspects of the economic situation which can be reflected by price relations;
  (b) Providing the information needed by the many users — Government, cooperatives and allied institutions, farmers, research workers — for the purpose of decision-making relating to economic activities.

12.2 Some Characteristics

A price must be related to the goods and services of strictly homogeneous contents. The price of a commodity may be defined as the amount of money paid by a buyer or received by a seller for a unit of commodity exchanged under given conditions.

When a price is quoted in a market, it is always accompanied either explicitly or implicitly by the specifications or characteristics of the

106 Agricultural Statistics: A Handbook for Developing Countries

commodity and the sales conditions. The more important characteristics are:

(i) The unit of measuring the quantity of the commodity for pricing purposes. It need not necessarily be of a single dimension. For example, although cotton may be quantified by a single dimension (in terms of weight), for eggs the use of one egg as a unit may not be adequate for temporal price comparisons as changes take place over time in the average weight per egg.

(ii) The specification of the commodity. Most important dimensions involved are variety, purity and size. Where varieties of a commodity are well established in the sense of botanical genetics, specification becomes easier but this is not as yet true for most countries nor for most commodities.

The purity of a commodity may be multi-dimensional. It includes the contents of other elements in the commodity such as moisture contents (e.g. rice in standard form is specified as containing 14% moisture), dirt and other particles.

The size is presumably an attribute to a variety, but in many cases is independent of the variety and it contributes one dimension of the specification attributes.

(iii) The conditioning of the commodity. This attribute refers specifically in which a commodity is prepared for sale — ways of packing or wrapping and the materials used for this purpose. It also relates to additives aimed at preserving or colouring the commodity and other non-processing types of additions which do not change the character of the commodity.

(iv) The time reference. Refers to the calendar time at which the agreement of the transaction aiming at the exchange of the commodity has been reached.

(v) The conditions of sale. This attribute is multi-dimensional and includes:
   (a) The location of market;
   (b) The place of delivery;
   (c) The time of delivery;
   (d) The time of payment;

(e) The quantity involved in a sale;
(f) The direct or indirect intervention of government.

When a price is quoted, all conditions of sale have to be specified explicitly or implicitly. These conditions are necessary for completing a transaction in a commodity.

12.3 Types of Agricultural Prices

The term "agricultural prices" is a general concept covering prices of all agricultural products and requisites for agricultural production, transacted at all stages of marketing. Two kinds of prices can be distinguished: (A) prices of agricultural products and (B) prices paid by farmers.

(A) *Prices of agricultural products*

This covers prices quoted for all farm produced agricultural products at all stages of marketing. Products include processed or unprocessed, originating directly from agricultural production. Distinctions between different types of prices are usually made on the stages of marketing. These are as follows:

(i) *Prices received by farmers.* These are prices quoted at the transaction of agricultural products in a market where farmers participate in their capacity as sellers of their own products. Generally the prices refer to prices received by the farmers at the farm-gate or first-point-of-sale nearest to the farm.

(ii) *Wholesale prices of agricultural products.* The wholesale market may be defined as a market situated somewhere between retail and producers' markets and usually handles a large quantity of sales other than sales for final consumption purposes. Those engaged in a wholesale market are usually well informed of the situation of the supply and demand condition of the commodity concerned; and therefore the price established there usually reflects the supply-demand condition of the commodity for the market.

The term wholesale price is also used to indicate the prices relating to transactions at all stages of marketing except the retail to consumer stage. Under this concept the prices received by farmers from wholesalers are also considered as wholesale prices but the distinction made above between these two types is important and useful.

(iii) *Retail prices of agricultural products.* These prices are established in the transaction in which the final consumers of the commodity participate as buyers.

(iv) *Export prices of agricultural products.* These are prices quoted in export market for commodities destined for delivery outside the customs boundary of the country. Although such markets may be considered as wholesale markets, they are generally treated separately. The valuation of the commodity at this stage is made as:

(a) f.o.b. (free on board) prices in exporting country;
(b) f.a.s. (free alongside ship) prices.

(v) *Import prices of agricultural products.* The price quoted at the importing point includes transportation, insurance, and other related charges involved in the transport of the commodity from the exporting country to the importing one — known as c.i.f. (cost, insurance and freight) prices. These prices exclude import duties.

(B) *Prices paid by farmers*

This concept is the counterpart of the prices received by farmers and covers all prices paid by farmers as they participate in the transaction of goods and services in their capacity as buyers.

There are two types of prices:

(i) Prices paid by farmers for the purchase of requisites of agricultural production;
(ii) Prices paid by farmers for their household consumption (in their capacity as consumers).

The requisites of agricultural production include all the material and services required in performing agricultural activities. These can conveniently be grouped into:

(a) Prices of materials used in current agricultural production — prices of raw materials used in the current process of agricultural production such as seeds, feed, fertilizers, pesticides, insecticides, fuel, oil;

(b) Prices of factor services — wage rates for farm labour, land rental rates, capital interest on fixed assets;
(c) Prices of investment goods — these cover the prices of equipment and machinery which are generally not fully consumed during one accounting year as well as all construction materials which are used for building up fixed assets on farms.

12.4 Collection

12.4.1 *General*

Whatever the uses made of price statistics, one of two main characteristics of such uses is generally always implied, i.e. the use involves either (i) comparisons of price quotations, or (ii) evaluation of a quantity of one or more products.

For (i) price quotations for given commodity specification and conditions of transaction are required over time in the same market or in different markets at the same stage of distribution, or for different specifications in one or more markets at the same stage of distribution. Other variants of price comparisons are also in use such as comparisons between different stages of distribution for the same commodity specifications, etc. The type of prices to be collected is determined by the type of comparisons to be served by the resulting price series.

For (ii) an average price per unit quantity of a commodity over a given period of time is generally required. In order to lead to accurate evaluation, the average price per unit should represent an average of all varieties of the commodities concerned and hence its calculation should take into account the quantities involved in the transactions through which the commodities are sold.

12.4.2 *Programme*

A programme of collection of prices should cover all the commodity specifications and conditions of sale required to serve the different purposes with adequate accuracy and at reasonable costs.

As a first step, one should have sufficient information on all marketing procedures and the different varieties of the agricultural commodities for which prices are to be collected – these can be secured by a proper survey. Such a survey should provide information on the location and size, the types of commodities dealt with, and the seasonal variations in the quantities sold in each market as well as the organization of sales in the different types of markets and for different types of commodities. The dates and duration of marketing in each locality should also be ascertained. These are essential in working out procedures to be used in the selection of markets and in the price collection methods. The survey should also lead to information on the variation pattern of prices of the commodities throughout the year with particular attention to seasonal patterns and changes over the marketing day. Data also should be collected on government taxation and subsidies.

The next step in setting up an agricultural price reporting programme is to plan the procedures and organizational set-up to be followed. These should be coordinated as far as possible with the programme for collecting non-agricultural prices. However, in the case of prices received[1] and of prices paid by farmers a more specialized and extensive administrative arrangement is required due to the distribution of the marketing channels over farms and village markets. Special interest is accorded to these prices because of their relationship to statistics of agricultural sector accounts and other social statistics that describe the welfare of the rural communities.

These prices can be collected from a sample of farmers. The farmers have to be interviewed at specified intervals to ascertain the prices they receive for a set of commodities under different marketing procedures and the quantities involved in each type of sale. The sample could be the same sample as for crop cuts or area measurements and some of the price data could be collected at the time of crop cuts or area measurements. However, further visits would be necessary for prices as prices have to be collected at specified intervals. The data collected by interview may not be reliable as these depend on what the farmer is able to recollect after an interval of time. The statistical staff can periodically check on such data by interviewing farmers on days of actual sales in local or central markets

---

[1] Generally known as agricultural producer prices.

where the farmers sell their produce. In fact in many countries the only consistent prices collected are from the markets.

12.4.3 *Sampling*

There are many practical and some theoretical difficulties that have to be overcome before sampling techniques can be applied to collection of prices. For example, it is not easy to define the sampling unit nor to construct a comprehensive and up-to-date sampling frame for the selection of the sample. Furthermore, a transaction where a price is quoted is a complex event with many dimensions. While a price quotation is an attribute of a transaction, the use of a transaction as a sampling unit in a single stage sampling design is not generally a practical procedure. Transactions take place in time and throughout the country. Transactions cannot be listed in advance. Accordingly, a recourse to area sampling procedures is inevitable.

As transactions take place in or near farms, in rural markets, in central and public markets and by wholesale and retail outlets, one may use these as primary sampling units. A list of farms or agricultural holders and lists of village markets, central and public markets, and of wholesalers and retailers dealing with agricultural products and requisites may be compiled to serve as a frame.

Stratification can be by administrative/geographical/marketing areas, and sub-stratification by size and type of commodities marketed. The first state sampling units could be holders (or farms), markets and other marketing outlets. The second and other subsequent stages of sampling should ultimately ensure the survey of representative samples, of transactions over time and in respect of the different commodities. It is not possible to consider all the details or varieties of sampling procedure. However, a number of points are of common interest and these are given below:

> (i) For comparison purposes a fixed sample that does not change over time might be suitable for compiling price series that are comparable over time and for other types of comparison. However, for evaluation purposes samples should generally be independent. Hence a compromise type of sampling is needed,

e.g. a changing sample with a fixed sub-sample would meet the situation.
(ii) The distributions of sampling units with respect to various types of measures of size or other attributes are generally very skew and hence in simple random sampling, adequate stratification of sampling units should be undertaken at least at the first, and where possible at the second stages of sampling.
(iii) The sample design may differ in detail and in its general characteristics according to the type of markets or marketing outlets being considered and also according to the type of prices as well as the commodities or groups of commodities for which prices are to be collected.

12.4.4 *Compilation from market reports and other similar sources*

The raw data on prices are obtained from many sources. Besides sampling or other enquiries already referred to, there are marketing reports and reports by agricultural and trade organizations of national or provincial coverage. In many types of market reports and other sources, information is presented in the form of ranges of prices in the transactions. The total volume of sales for the day may also be given. Sometimes the "modal price" is given. Such information should be processed and published. The frequency of publication depends on many factors. For some commodities when daily variations are quite appreciable these should be published weekly or at least monthly. Where there are no significant variations, monthly publications would suffice.

12.5 Users

It is very important that those charged with the collection of price statistics should regularly discuss their work with potential users. Users could be the various Government Departments as well as outside sources such as Chambers of Commerce, trade organizations, and farmers' organizations.

CHAPTER 13

Statistics for Agricultural Planning

13.1 Inclusion in Existing Data Collection

In the process of collecting current agricultural statistics in developing countries, data for planning could be collected by inserting additional questions in the relevant questionnaires without much further effort and expenditure. It may not be possible to cover all the required data for planning by this method, in which case, special surveys or enquiries will have to be conducted.

Agricultural statistics for planning cover a wide range of topics, some of the important ones are given below:
   (i) Land utilization;
   (ii) Irrigation;
   (iii) Fertilizer usage;
   (iv) Employment in agriculture;
   (v) Agricultural power and machinery;
   (vi) Agricultural credit;
   (vii) Market intelligence;
   (viii) Costs.

13.2 Land Utilization

A decennial census provides a country with data on land utilization under the broad categories arable land, land under permanent crops, land under permanent meadows and pastures, wood or forest land, and "all other land" by small administrative divisions. For serious studies on land utilization, these data alone will not be sufficient. The different categories of land have to be located. Hence maps of the country by its regions depicting the various categories should be prepared at the first instance. Some developing countries do have such maps, with details on soil types, rainfall pattern, etc.

Such maps with the supporting tables would give general ideas for broad changes in the utilization of land; for example a part of the land under wood or forest can be made arable land. But equally important is to know how the land, already in use, is profitably utilized.

For such purposes data on crops including crop intensity and cultural practices, use of irrigation, fertilizer, etc. would be needed season by season. This means that when data on crops such as area and yield are collected, care is taken to include the breakdown needed for each crop by varieties of seed used, by cultural practices adopted, by types of irrigation, etc. When these estimates are made by sample surveys, sample size should be increased so that the errors are tolerable for the sub-groupings. When assessing the benefits of some of these practices in land utilization, care should be taken in the analysis of the data of interaction between some of the practices.

Items for data collection on land utilization would vary from country to country and it is necessary for each country to prepare its own items. The items should be settled by the authorities in charge of agricultural statistics in consultation with those dealing with land use, agriculture, irrigation and planning.

Here is a land use table for an Asian country to illustrate the type of data collected:

13.3   Irrigation

Water is a limiting factor in most types of agriculture and therefore it should not be wasted and maximum use made of it. Much of the water

	1962 (estimated)			1975 (proposed)		
	Million ha	Percent of Total area	Percent of Sub-Total	Million ha	Percent of sub-total	
Land and inland water: total area	51.40	100		51.40		
Area not used for agriculture	14.80	29		14.38		
Area uses for agriculture	36.60	71		37.02		
– forests	28.00	54		26.40		
– permanent pastures	–			–		
– arable land and land under permanent crops: total	8.60	17	100	10.62		
– irrigated	1.69	3	20	2.24		
– not irrigated	6.91	14	80	8.38		
Area harvested: total	8.27	16	100	10.37	100	
– irrigated	1.79	3	22	2.41	23	
– not irrigated	6.48	13	78	7.96	77	
Cropping intensity: total		96		98		
– irrigated		106		108		
– not irrigated		94		95		

Cropping intensity is calculated as a percentage of area harvested to the physical land area. For instance, the cropping intensity for 1962 of irrigated land is

$$\frac{\text{Harvested area}}{\text{Physical area}} \times 100 = \frac{1.79}{1.69} \times 100 = 106$$

The area harvested for a year is the sum of the area harvested for each season during the year and can therefore exceed the physical area.

available in irrigation schemes is under-utilized for lack of attention to land levelling, field distribution, drainage systems and regulatory services.

For purposes of agriculture, data on irrigation are collected by the amount of land benefited. Data could be collected by modes of irrigation, an example of which is given below:

	Area under irrigation (ha)
I Surface water	
(a) Gravity	
(i) Storage	–
(ii) Diversion	–
(b) Pumping	–
II Groundwater	
(a) Deep wells	–
(b) Shallow wells	–

In some Asian countries the tabulations of crop acreages distinguish the area of each crop in three parts, according to whether it is watered by major irrigation, by minor irrigation or is solely fed by rain.

Besides such data, data are needed on the utilization of available water. It may be that in a particular season that though water was available for a certain acreage of a crop, a smaller extent was cultivated. Thus volume of water available and water utilized for the several crops have to be collected to give some guidance for better utilization in the future. Further, data on the volume of water needed for the different cultural practices of the several crops should be collected. Such data along with production figures would enable planners and agriculturalists to decide on the merits of the different cultural practices *vis-à-vis* production and use of water.

In countries where there are sizeable areas under new irrigation projects, data should be made available by each project. Such data are needed for project appraisals, evaluation and for planning for the future.

Data for drainage and flood control are needed as for irrigation.

13.4 Fertilizer

Data on fertilizers and soil dressings provide the area treated for each crop. For inorganic fertilizers, in addition, data on quantities used are needed.

Data on chemical inorganic fertilizers used are given by the main plant nutrients. These are classified as:

    (i)   Nitrogeneous fertilizers;
    (ii)   Phosphate fertilizers;
    (iii) Potash fertilizers;
    (iv) Mixed, compound and complex inorganic fertilizers.

Areas treated by organic manures such as farmyard manure, compost, green manure and seaweed are sometimes collected. It is generally difficult to collect such data while it is comparatively easier to collect data on chemical inorganic manures due to the fact that farmers purchase specific quantities of fertilizers each season and therefore they are in a position to supply the data of the distribution of the fertilizers applied among the various crops. This is not the case for organic manures which are generally collected *ad hoc*.

Data on the use of chemical fertilizers are very important as these constitute the most important single weapon on the food production battle.

Analogous considerations apply to the use of other agricultural chemicals, for example, in plant protection and weed control. Hence data on pesticide too have to be collected in addition to those on fertilizers.

13.5 Employment in Agriculture

Basic data on employment in agriculture are collected in the decennial censuses. These data should be kept up-to-date by periodical sample surveys. It is customary in some Asian countries in sample surveys for crop cuts, data on employment in agriculture and related data are collected on a sub-sample basis.

Data can be collected by (a) holder and unpaid members of his household; (b) participants in cooperative, collective and communal holdings; and (c) persons working for pay in the holding. Each of these can be sub-divided into (a) permanent, temporary and occasional workers and (b) males and females.

To obtain data on employment in agriculture, further items can be added to the surveys requesting information on number of man-hours spent on (a) agricultural and (b) non-agricultural work. As agricultural work is seasonal, such data should be collected at different seasons.

It is useful in these surveys to collect data on wages paid to agricultural labour.

13.6 Agricultural Power and Machinery

For good planning, it is necessary to know whether the various agricultural operations are performed by human power, animal power and mechanical power. In some countries, acute shortage is felt particularly during the planting and harvesting periods of some or all of human, animal and mechanical power. In countries with plenty of human labour and shortage of foreign exchange to import machinery, agricultural schemes are geared to labour intensive methods. It is therefore necessary to collect data on the various items mentioned on a seasonal basis. A convenient way to collect such data would be on a sub-sampling basis in sample surveys conducted seasonally to determine yields or areas. It is not necessary to collect these data every year as changes may not be significant as to warrant extra work on the statistical staff on a regular basis.

Whenever such data are collected, data on general transport facilities available for agriculture can also be collected. These would include transport of inputs to the farms and transport of products out of the farm. The mode of transport has to be specifically stated, such as:

by foot;
by pack animal;
by bicycle;
by motor vehicle;
by train.

13.7 Agricultural Credit

It is a well-known fact that in developing countries, small farmers are chronically in debt. It is sometimes so bad that loans are taken from traders on the security of the standing crop which means that the farmers never get the full price for their produce. Thus even for cultivation in the traditional methods the small farmers need credit. If, however, in the wake of the green revolution, cultural practices requiring additional inputs are adopted, there will be a need for still further credit.

To assist the farmers with the needed credit, data on past credits granted and amounts repaid are necessary to give guidance. Collection of such data should be placed in the list of current agricultural statistics collected.

Further data on taxes collected from agriculture and subsidies paid are necessary to have a clear picture on the credit needs.

Some countries have found it prudent to make available part of the credit to farmers in kind to avoid any temptation for misuse of cash loans. These take the form of providing some of the inputs needed, like fertilizers and seed. Such facts have to be taken into consideration when data are collected.

13.8 Market Intelligence

Statistics under this heading should include those of the quantities of cash crops marketed through major marketing channels and prices at which the produce changes hands. Further, as many developing countries suffer from inefficient marketing, statistics relevant to marketing efficiency and for improvement of the marketing system should be made available.

Many developing countries have full-fledged Marketing Departments and in some, these organizations are part of other big organizations like Cooperatives or Agricultural Departments. The marketing organization is spread throughout the country with its own field officers. The organization collects data on marketed quantities and prices at which produce changes hands mainly in the areas which have surplus produce for sale.

In the existing system the first step to be taken is to have better coverage. The areas not covered can be known from field officers or from data collected from other sources (like production data from Statistics or Agricultural Departments). The coverage should be extended to more crops, gradually. There should be greater dissemination of information to the public. This helps particularly the small producer to get better prices for his produce. For example, in Sri Lanka, the national radio broadcasts every day (mid-day) the day's prices of most of the agricultural products. After this came into existence, there have not been many complaints from small producers about the prices received by them for their produce.

Many marketing experts from developed countries while working in developing countries have remarked that the "green revolution" would be a greater success if the marketing system in these countries were more efficient. Lack of infra-structure, collecting centres, storage facilities, and reasonable prices for the produce are some of the factors contributing to the existing state of things. Having these in mind, FAO introduced a scheme of collecting data village wise which would reveal such gaps.

Loss of produce through inefficient handling, and processing and at storage places (mainly by rats) needs serious consideration by marketing organizations. Reliable data are not available on such losses but the small amount of data available indicate that the losses are serious. Steps will have to be taken to remedy the situation for which reliable data are needed. This is one direction where data collection will have to be built up.

Surveys would be necessary for collecting data on special aspects of marketing. For such purposes, it would be necessary to maintain up-to-date frames of marketing and collecting centres, storage places, lists of growers of particular produce (mainly for marketing), processing plants and others.

13.9 Cost of Production

Data on cost of production of many agricultural products are requested from statisticians by agricultural economists and planners. Hence data collection for this purpose has become a routine collection in some countries. A convenient way of collecting these data would be from a sub-sample of farmers when sample surveys are conducted for crop cut or area measurement. Otherwise special sample surveys would have to be conducted for this purpose.

In the case of crops, data on costs are collected against the area

cultivated under the following items:
  (a) Cultivation operations;
  (b) Harvesting operations;
  (c) Materials and supplies.

Data are also collected on the value of produce thus enabling a profit and loss account to be prepared.

If data are needed by sub-divisions of these items, questionnaires used in the surveys should provide for these sub-items.

A typical form used in an African country is given in the Appendix. Cultivation costs are collected by the type of operation and all the operations have a breakdown by family and hired labour. If animal and mechanical power are used, the form has to be suitably amended.

On the income side value of the produce and that of any subsidiary produce are included.

Data published are mostly given per unit of land and in some cases per unit of weight of produce.

CHAPTER 14

Staff and Organization

14.1 Introduction

Different countries have different organizational set-ups for the compilation of agricultural statistics. Central Statistics Offices (CSO) are in charge of the subject in some countries, in some others the subject comes under the purview of the Ministry of Agriculture (M/A) and there is a third group where there is dual control (by CSO and M/A). So long as any system works, there is no need for a change.

Whatever the set-up, the immediate responsibility lies with the statistician in charge of the agricultural statistics division. Though he receives directions from higher authorities (like Director, Director/General, Permanent Secretary), it is he who plans and executes the work with his office and field staff.

14.2 Staff

14.2.1 *Office staff*

The office staff would work mainly at the Head Office and in some cases in regional/provincial/district offices. The officers would be in two categories (a) mainly technical, (b) others/non-technical. The technical staff would consist of statisticians, assistant statisticians, statistical officers and statistical investigators who would be the supporting staff for the statistician-in-charge for all his work. The staff should be sufficient for all the operations inclusive of editorial scrutiny and tabulation. If tabulation is done by the data processing unit located outside the agricultural statistics division, some members of this unit should be allocated to handle

Staff and Organization

all data processing in agricultural statistics.

For organizing and executing the work at all levels in the office ensuring uniformity and continuity and without disruption, officers should work in groups of two or more so that the absence of one officer should not put the system out of gear. The other officer or officers of the group should be able to carry on.

Non-technical staff would consist of clerks, typists and others who generally belong to a central service and are liable to periodical transfers.

14.2.2 Field staff

Field staff would consist of various categories of officers from field enumerators to field supervisors. Field staff should be stationed in the various parts of the country which has many advantages. For such a system it is useful to recruit the junior field staff locally. The advantages of the system are economy in travelling costs, no language barrier, benefit of knowledge of local agricultural conditions and goodwill and cooperation of farmers. A serious disadvantage, however, has to be reckoned when the local officer does not perform his duties satisfactorily and the supervisor who is invariably an outsider finds himself in a difficult position to control him. However, the advantages would outweigh the disadvantages.

In a few countries they have mobile field staff. These field officers are moved from place to place in teams depending on the statistical needs of the areas and when they have no field work, they are in headquarters. This system works only when few surveys are conducted in a country but if statistics are collected on a regular basis, there is no alternative but to station staff in the various parts of the country.

14.3 Selection and Training

14.3.1 Selection

Various grades of officers need various prescribed educational standards to be eligible for selection. Besides the educational achievements, certain qualities are needed to make a success in agricultural statistics, such as:

(a) Capable of field work under village conditions;
(b) Should possess tact and respect for local social courtesies, and friendly to farmers;
(c) Reasonable good health for travel and for occasional living in villages;
(d) Intellectual honesty and capacity for looking after his work without much supervision.

It would thus be seen that the success of an officer handling agricultural statistics depends not only on his technical skill but also on his capability of handling rural people.

14.3.2 Training

Training is required at a number of levels depending on the level of the officers:
  (i) For enumerators;
  (ii) For supervisors;
  (iii) For statisticians.

Training can be broadly divided into (a) basic and (b) in-service. Basic training is given to staff recruited in the junior grades immediately after recruitment and before being given responsible work. In-service training is given on a regular basis for appraising the officers in service of modifications in the methodology and collection of data or for special surveys or enquiries.

Training of enumerators and supervisors is the most important training programme in any country. Except for basic theoretical knowledge, the training should be geared to work in the field. Therefore, the officers conducting the courses should mainly be drawn from those who have actually handled the work in the field in their earlier years and are still in contact with the field. The training should include field trips which should not only demonstrate the methods to be used but also get the trainees involved in the work by actual participation. For example, if the training is for census taking, it should include the full participation of the trainees in a pilot census.

In any training course, the skills needed to be a good field officer should be imparted. If the officers are receiving the instruction for the first

time, then it can be a detailed course otherwise it should be brief so as to fit into the course given for other purposes. Some of these are given below:
- (i) Simple arithmetic calculations including fractions, decimals, proportions and percentages;
- (ii) Calculations of distances and areas; use of conversion tables;
- (iii) Basic ideas on sampling and use of random tables;
- (iv) Use of maps for locating lands and sketching them;
- (v) Use of simple field instruments to measure angles and lengths and for weighing produce;
- (vi) General ideas of the agricultural practices likely to be met with;
- (vii) Basic skills in approach and interviewing;
- (viii) General methodology in surveys;
- (ix) Drafting reports;
- (x) Administrative matters to be dealt with in the course of his duties.

The training courses, besides providing the training needed by the junior staff, also provides the senior staff to spot out those officers who display special skills in certain fields. Some officers may excel in some type of field work, others in office work and a third group in supervisory capacities. Channelling the officers to suit their inclinations would be fruitful both to the officers and to the authorities.

Supervisors should attend all the training courses given to the enumerators. It is only then that they would be conversant with the work they have to supervise. Further training is needed for the supervisors. Being better educated and more experienced than enumerators, supervisors should be able to receive more technical instructions. The statisticians conducting the courses should impart to the supervisors their experiences particularly in the direction where enumerators are likely to slip and where attention has to be mostly concentrated. They should also be instructed not to act as mere post offices passing questionnaires between the enumerators and the head office.

The training of statisticians is a long process. In the early stages the statistician should fill in the gaps of his theoretical knowledge by attending lectures on such courses in a local University, while at the same time working in close association with a senior statistician who has to put him on the way. To gain experience in the field the statistcan should work along with the enumerators for a month or two and likewise for a reasonable period with supervisors. In the office he should familiarize himself with the methodology and techniques used for the several surveys conducted and data collected as well as with editing, analysis and presentation of these data in the form of reports. A short course in management is essential as eventually he will have to manage a large office and field staff. After a few years as a statistician, it is worth sending him abroad for a suitable course in a foreign University (specialized in agricultural statistics) at a post-graduate level, to enable him to become a full-fledged statistician.

14.4 Staff Management

It is important for staff management and control that the contact between the head of a statistics organization and the staff at various levels and in the various geographical locations should be effective. A way of achieving this is by regular meetings with the field staff. In a country with provinces and districts, (a) all field staff below district level could meet at regular intervals at the district headquarters where the statistician of the district presides, (b) all district officers could meet at provincial headquarters presided over by the provincial statistician, and (c) all provincial officers could meet at headquarters where the Director/General or Director of the Statistics Organization presides. These meetings could deal with all technical and administrative matters between the field staff and the officers controlling them. Dialogue and personal contacts are essential for smooth functioning of organizations. Besides these meetings, senior officers could make routine field inspections where some of the outstanding problems could be sorted out. Any misunderstanding or delays in correspondence could be straightened out at such meetings.

The central organization should have complete disciplinary control over its entire staff. Instructions given should be strictly complied with. Mistakes should be pointed out and corrected. If repeated, serious action has to be taken. Equally, good work should be appreciated. In the exercise of control, some flexibility is needed depending on the type of subordinate officers. For example, if the superior is very strict, it is admitted not many mistakes would be made — however, cooperation from the subordinates would not be spontaneous. On the other hand if with a superior who is good to his officers, cooperation would be forthcoming but a minority would exploit the situation. So that one uniform way of control is not feasible; control need not be rigid for smart officers but has to be rigid for others.

Enquiries which warrant the stay of the enumerators in the villages for a length of time (some weeks or months) bring in special problems. Frequent visits by the enumerators to the farmers' homes would be made. Some farmers would not be happy on such a situation particularly when the visits are made in his absence and where there are young females in the household. In such cases, it would be wise for the supervisor to meet the farmers at the outset preferably in the company of some village officer or leader and explain the purpose of the enquiry and the need of frequent visits by the enumerator. The enumerator on the other hand should be advised not to be too involved with his informants and that his visits should be made preferably when the farmer is at home.

Instances arise when an enumerator reports that a farmer refuses to cooperate to make available the data needed. In such cases the supervisor or a more senior officer would have to meet the farmer and talk to him tactfully to get the data. Sometimes such action tends to undermine the position of the enumerator in the eyes of the public. This can be overcome if the supervisor allows the enumerator to record the data instead of recording it himself.

Though much can be said on staff management and control, it is mostly by experience that one learns and makes a success of them.

14.5 Field Organization

14.5.1 *Set-up*

Many developing countries have permanent field organizations for the collection of agricultural statistics, some through the length and breadth of the country and others for important agricultural regions of the country. Taking into consideration the availability of technical personnel, funds and volume of work perhaps it is not necessary at present to have field officers for every village. Field officers in charge of a group of villages or a sub-district would be sufficient at the lowest level.

These sub-district officers are supervised by a district officer and district officers in turn by a provincial/regional officer and finally the provincial/regional officers by the Central Statistics Organization.

There are variations to this type of set-up. When there are not sufficient statistical sub-district officers, agricultural field officers at this level are used to cover the duties. For special surveys additional enumerators are appointed who work with the sub-district officers.

Field offices. There are no proper offices at sub-district level — a room at the residence of the officer is sometimes used to meet the farmers. Invariably the officer meets the farmers in the field or in the farmers' residence.

Proper offices are provided at the district level. Documents received from the field are scrutinized in these offices and despatched to provincial/regional offices or to the head office depending on the instructions given. Documents from the opposite direction also pass through these offices. These offices are the headquarters for the sub-district officers and regular meetings are held here with the district officer. Farmers and others interested in agricultural statistics call at these offices for any needed data.

The district offices generally maintain the forms, questionnaires and equipment needed for the entire field staff of the district. These offices also maintain the motor vehicles needed for staff travel.

The provincial/regional and head offices are used only occasionally by the sub-district field officers: These are meant for dealings with officers at the district level and for contact with the head office.

14.5.2 *Public relations*

The statistical organization should foster good relations with other government departments and agencies at all levels. The important departments are those of Agricultural and Local/Home Affairs. Other departments include Labour, Cooperatives, Commerce and Industry and statutory organizations for marketing of agricultural products and for research (coconut, tea, rubber, palm oil for Research Institutes). Enquiries relating to subjects of any of these bodies should be organized preferably with their prior concurrence. The liaison with the organizations is vital in the field. In the day-to-day affairs, the statistical field staff will have to work in close relationship with the field officers of the various organizations and particularly with the agricultural extension officers.

It is useful to keep the interested organizations informed periodically of the current activities of the statistical organization and particularly when there are special surveys. A timetable should be provided for all surveys to those directly involved.

At the commencement of any enquiry, it is useful either to make use of some regular meeting of farmers or to hold a special meeting explaining the purpose of the enquiry and to stress the benefit to the development plans of the country if the data sought are given. People whom the farmers respect should be approached to address them seeking their cooperation and assuring them that the data collected would be treated as confidential and would not be used for tax purposes. All contacts with farmers and their families should be done with utmost patience and courtesy. For all enquiries the farmers should be contacted at the time and place most convenient to them. Many of the farmers are busy in the field in the mornings and therefore any form of enquiry should be arranged for the afternoon or evening. If field measurements are involved then these should preferably be done when the farmer is in the field. When an enquiry or a field operation is completed, the enumerator/field officer should make it a point to thank those who helped him in his work.

14.5.3 *Regular work*

Field work would consist broadly of (i) regular/routine work and (ii) special enquiries and surveys.

Regular work would consist of data collection on crops and livestock periodically (monthly, quarterly, annually) and despatching the returns on due dates to the senior officer. For efficient handling of routine work, proper time tables should be provided to all field officers and the Head/Provincial/District offices should see that the returns are received on the due dates. Supervisory staff should provide for handling of urgent work in the absence on leave or otherwise of any of the field officers. The field officers should arrange their field trips taking into account the convenience of the farmers and keeping to the timetable provided by the supervisor. Provision should be made for visiting the field or interviewing the farmer on more than one occasion for a single job. With all the instructions and care taken in such matters, delays inevitably take place. It is the writer's experience in Indonesia that not all monthly returns of paddy and other food crops are received in time at the central organization. Generally 50% of the returns are received on time, about 25% within a reasonable amount of time after the due date and the balance of 25% need chasing from the field. The central statistical organization has thus to provide for a certain amount of delay from the field. What can be done is to reduce this to a minimum by greater supervision and prompt follow-up actions.

The forms used for routine work need periodical revision in the light of the experience in the field. At the regular meetings between the senior staff and the field officers the procedure in filling the forms should be reviewed. It is sometimes found that not all the data collected are made use of; collecting such data may be dispensed with and the items for data collection could be reduced. The frequencies of collecting data on some items have also to be reviewed.

Field officers in many countries are transferable sometimes to any part of the country and other times to some other area of the province/region. The question which is often asked is whether this helps for better efficiency. A field officer to give his best, would take a certain amount of time to familiarize himself with the area. Hence transfers after short spells in an area would not help for efficiency; in fact it will have the contrary effect. On the other hand, a long stay helps in many ways but carries the hazard that the officer assuming things do not change much, may repeat some data, without visiting the area. A compromise is the answer; a reasonable stay of three to five years would be in the interests of efficiency.

14.5.4 *Special enquiries/surveys*

If special enquiries/surveys are conducted by the regular field staff then what has been described in section 14.5.3 applies here. When the enquiries are conducted mainly by temporary staff (enumerators) some additional steps will have to be taken. These include grouping the temporary enumerators so that each group comes under a permanent field officer or a permanent enumerator for close supervision and control. Any little problem in the field has to be promptly dealt with and any discrepancies rectified without delay as the enumerators would not be long with the organization. Completed questionnaires should be periodically collected from the enumerators and scrutinized. There should be proper inventories of questionnaires, equipments etc. handed over to the enumerators and at the time of separation these should be checked and taken over. Precaution should be taken not to make the final payments to the enumerators till all work on the enquiry/survey is over and all questionnaires and equipment are returned.

Experience has shown that for special enquiries/surveys it is preferable to use permanent officers than temporary enumerators. If permanent officers of the statistical organization are not sufficient, officers from other organizations can be used. Field officers of the Agricultural Department, cooperative officers, marketing officers, teachers on school vacations and others can be used. The quality of the work is generally better with these officers, sense of responsibility better and they are more amenable to discipline than with temporary staff.

For all enquiries, surveys and routine data collection, the field staff should be provided with a manual of instructions which embody all important instructions needed by them for carrying out their duties (technical as well as non-technical).

Appendices

Typical form used in an African country for seasonal collection at Sub-district level

(Sample Survey for data on cost of production)

Crop: ....   Variety: ....
Season: ....

A. Province: ....   Name of tenant: ....
   District: ....   Extent: .... ha
   Sub-district: ....

   Is the land operated alone or in partnership or on rental: ....
   If in partnership, give details: ....
   If on rental, amount of rent/season: ....

B. (i) Preparation of land: No. of man days   Family   Hired
      (a) Sowing operations: ....
      (b) Providing irrigation and
          other maintenance operations: ....
      (c) Use of fertilizer: ....
      (d) Use of pesticide: ....

   (ii) Harvesting operations: ....

   (iii) Any other operation: ....

   Labour rate man/day: ....

C. (i) Amount of ⎫ Seed   .... kg
      Cost of   ⎭   .... kg
   (ii) Name of   ⎫   .... kg
       Amount of ⎬ fertilizer .... kg
       Cost of   ⎭   .... kg

(Reference Chapter 5)
Typical form used in an Asian Country for monthly collection at sub-district level
(Reporting system – mainly data on cropped areas)

Province: ....
District: ....
Sub-district: ....

Form No.: ....
Month: ....
Year: ....

		Area in hectares (ha) (Single Crop)								Area in ha (Mixed Crop)							
	Name of Commodity	Planted area (end of previous month)	Harvested area fully ripe (during month)	Harvested area not fully mature (during month)	Area damaged* Total damaged area	Area damaged* Harvested Area (ha)	Area damaged* Harvested Production (100kg)	Newly planted area (during month)	Planted area (end of reporting month)	Planted area (end of previous month)	Harvested area fully ripe (during month)	Harvested area not fully mature (during month)	Area damaged* Total damaged area	Area damaged* Harvested Area (ha)	Area damaged* Harvested Production (100kg)	Newly planted area (during month)	Planted area (end of reporting month)
(1)	(2)	(3)	(4)	(5)	(6)	(7)	(8)	(9)	(10)	(11)	(12)	(13)	(14)	(15)	(16)	(17)	(18)
1	Paddy																
	(i) Irrigated																
	(ii) Non-irrigated																
2	Maize																
3	Cassava																
4	Groundnuts																
5	Soyabeans																

* by floods, drought, catastrophy, pests etc.
Col. (3) – col. (4) – col. (5) – col. (6) + col. (9) = col. (10)
Col. (11) – col. (12) – col. (13) – col. (14) + col. (17) = col. (18)

Signed: ....
Statistics/Agricultural Officer – Sub-District

(iii) Name of ⎫ .... kg
     Amount of ⎬ pesticide .... kg
     Cost of ⎭ .... kg

    Cost of other inputs ....
    ....
    ....

D. Amount of produce sold: 100kg
   Amount of produce retained for consumption: 100kg
   Amount of produce retained for seed: .... kg
   Amount of produce given as share if cultivated in partnership:.......... 100kg

   Produce sold at farm/elsewhere
     If elsewhere cost of marketing: .... per 100kg
   Amount obtained by sale of primary produce:...................
   Amount obtained by sale of secondary produce: ....

Relevant Publications

A. Hunt, K. E. *Agricultural Statistics for Developing Countries*, Oxford, The Institute of Agrarian Affairs, 1969.
   Theodore, G. *Statistiques Agricoles dans les Pays en voie de Developpment*, Centre European de Formation des Statisticiens Economistes des Pays en voie de Developpment, 1966.
— United States Department of Agriculture. *Statistical Reporting Service of the U.S. Dept. of Agriculture (Scope, Methods)*, Washington, 1964, Miscellaneous Publication, No. 967.

For comprehensive accounts of agricultural statistical services.

B. Cochran, William G. *Sampling Techniques*, New York, Wiley, 1965.
   Sukhatme, P. V. *Sampling Theory of Surveys with Applications*, Indian Society of Agricultural Statistics, New Delhi and Iowa State University Press, Ames, Iowa, 1963.
   Yates, Frank. *Sampling Methods for Censuses and Surveys*, London, Charles Griffin, 1960.

As textbooks on sampling methods.

C. FAO: *Lectures on Rural African Surveys*, Rome, 1957.
   FAO: *Handbook on Agricultural Sample Surveys in Africa (Principles and Examples)*, Rome, 1957.
   Koshal, R. S. A review of work done on development of sample surveys for the estimation of agricultural production in United Arab Republic, 1962.
   Narain, R. D. *Methods of Collecting Current Agricultural Statistics*, Rome, FAO, 1955.
   Panse, V. G. *Estimation of Crop Yields*, Rome, FAO, 1954.
   Sastry, K. V. R. *Report on the Improvement of Coconut Statistics in Countries of Asia and the Far East (1960-65)*, Rome, FAO, 1965, EPTA, No. 2093.
   Zarkovich, S. S. *Estimation of Areas in Agricultural Statistics*, Rome, FAO, 1965. *The Quality of Statistical Data*, Rome, FAO, 1966.

For experience in agricultural statistical work in the different fields as specified in their titles, in developing countries.

Relevant Publications

D. FAO: *1970 World Census of Agriculture.*
   Regional Programme for Asia and Far East 1967.
   Regional Programme for Africa 1967.
   Regional Programme for Near East 1967.
   Rome, FAO.

   Khamis, S. H. and Alonzo, D. C. Highlights of the 1980 World Census of Agriculture, *Monthly Bulletin of Agricultural Economics and Statistics*, Rome, FAO, December 1975.

   Panse, V. G. *Some Problems of Agricultural Census Taking, with Special Reference to Developing Countries*, Rome, FAO, 1966.

   Zarkovich, S. S. *Sampling Methods and Censuses*, Rome, FAO, 1965.

   On agricultural census.

E. FAO: Food Balance Sheets 1964-66, Rome, 1971.

   On all aspects of the preparation of national food supply/utilization accounts with accounts for most countries of the world.

F. Houseman, E. E. and Huddleston, H. F. Forecasting and estimating crop yields from plant measurements, *Monthly Bulletin of Agricultural Economics and Statistics*, Rome, October 1966.

   Sanderson, Fred H. *Methods of Crop Forecasting*, Harvard University Press, 1954.

   On crop yield forecasting.

G. FAO: *Statistics of Livestock Numbers and Livestock Products in Asia and Far East*, AGS: FE/3/70-71, Rome, 1970.

   On estimation of livestock numbers and livestock products.

H. FAO: *Report of the Regional Seminar on Agricultural Producer Price Statistics* (TA 2169), Rome, 1966.

   FAO: *Statistics of Agricultural Prices* (FAO-ECE-Conf. Eur. Stats/AP3), Rome, 1968.

   On agricultural prices.

Index

When a subject is dealt with in the whole of a section,
only the first page of the section is entered

Accuracy   21, 34, 35, 66, 86, 87, 94, 98, 109
Administrative areas/units   19, 21, 32, 34, 39, 53, 67, 101, 111, 113
Aerial photographs   56
Agricultural power and machinery   86, 115, 118
Agricultural production, general ideas   14, 15, 16, 74, 77, 86
  For other aspects see Areas; Crop yields; Forecasting; Livestock and Livestock products.
Agricultural credit   113, 118
Agricultural statistics
  basic and current   2
  definition   1
  development   3
  macro and micro   2
Areas
  cadastral/non-cadastral   8, 12, 20, 35, 36, 40
  cropped   33, 35, 40
  gross/net   38
  measurements   12, 40
  measurements, of angles   41
  measurements, of distances   41
  measurements by sampling   35, 38
  reporting   see Crop reporting

Bias
  in census   94
  in estimation   24, 27
  in forecasting   49
  in selection   103

Census
  of agriculture   31, 53, 84

complete enumeration   84, 90
concepts and definition   84, 88
evaluation   94
objectives   84, 85
planning and procedure   90
post enumeration   95
sampling   84, 90
scope and coverage   86
tabulation   93
time reference   89
Checks on field work   37, 126
Correlation   24
Costs   13, 21, 38
  of production   90, 96, 120
Coverage   37, 64, 75
Crop cuts/crop cutting plots   11, 13, 19, 24, 81, 98
  harvesting and weighing   26
  size and shape   25
Crop estimation   15, 16
Crop forecasting   46
  areas   46
  condition-yield relation   49
  production   46
  yields   47
  yields, objective method   50
  yields, subjective method   48
Cropping intensity   114, 115
Crops
  permanent   11, 14, 33, 44, 113
  temporary   11, 14, 33
Crop yields   5, 7, 11, 15, 19, 28, 31, 47

Development
  methodology   6
  priorities   7
  selection of priorities   9
Domains of study   21, 102

137

Index

Egypt 28, 38
Employment in agriculture 86, 89, 113, 117
Enumeration, unit 7
Enumerator 4, 12, 57, 67, 92, 103, 123, 124, 127, 128
Estimation of mean yield 26

FAO 2, 46, 55, 61, 69, 72, 80, 89, 93, 120
Fertilizer, usage 85, 87, 113, 117
Field work *see* Staff and organization
Food consumption surveys 66, 67, 75
Food Supply/Utilization Accounts 13, 64, 74
(Food Balance Sheets)
 concepts and definitions 77
 coverage, commodity 75
 example 81
Forecasting *see* Crop forecasting
Forms 130
Frames 11, 18, 99, 111
 for crop surveys 20
 for livestock surveys 56

Holder 92, 111, 117
Holdings 12, 15, 40, 86, 89
Horticulture 12
Households, agricultural 20, 100

Index numbers
 agricultural production 64, 69, 71
 example 72
 food production 64, 71
 formula 70
 weights 70
India 11, 14, 17, 24, 38
Indonesia 17, 35, 47, 88, 98, 130
Irrigation 85, 87, 113, 116

Koshal 29

Land
 tenure 2, 85
 utilization 2, 85, 113
Laspeyrés-formula 70
Livestock 14, 53, 86
 census 53
 inventory 53
 nomadic 57
 sample surveys 54, 56
Livestock products 14, 60, 86
 meat 60
 meat weight 61
 milk, milk products 63
 offals 61
 poultry, eggs 63
 reporting 62
 sample surveys 66
 slaughter fats 61

Malaysia 14, 98
Maps
 areas from 37, 39, 40
 use as frames 100
Machinery, and power in agriculture 86, 113, 118
Market intelligence 113, 119
Mauritius 17
Mean
 arithmetic 26
 estimate 26
 weighted 28
Methodology
 for census 84
 of development 6
 for surveys 97
Ministry/Department of Agriculture 3, 8, 52, 56, 119, 122, 128
Mixed cropping 42

Nadarajah 57
National Farm Surveys 2
Neyman 22

Objective methods 3, 6, 10, 12, 19, 37, 96
Objectives of census/survey 85, 97
Optimum allocation
 between strata 22

Index

Organized/non-organized sectors 8, 20
Organization *see under* Staff and organization

Planning, agricultural 1, 3, 113
Population, statistical 160
Precision 21, 28
Prices 105
  characteristics 105
  collection 109
  producer 110
  sampling 111
  types 107
  wholesale, retail, export, import 107, 108
Probability of selection
  equal 24, 27
  proportional to size 24
Public relations 128
Purposive selection 68

Questionnaires 37, 59, 91, 93, 113, 121, 128

Random numbers 23
Ratio estimate 28
Reporting, levels of 30
Reporting system 3, 33, 55, 62
Reporting unit 34, 36, 55
Reports, condition 48

Sampling
  application 11, 98
  crop yields 19
  crop yields, examples 28, 31
  errors 11, 19, 90, 94, 114
  fraction, constant and varying 102
  livestock 54, 56
  livestock products 66
  methods 38, 97
  multi-stage 23, 100
  nomadic livestock 57
  prices 110, 111

  process, for crop yields 19
  techniques 3, 34
  units 19, 21, 99
  use of 11, 19, 97
Sastry 31
Selection with probability proportional to size *see* Probability of selection
Somalia 57
Staff and organization 122
  field organization 127
  field staff 123
  office staff 122
  public relations 128
  selection, staff 123
  training, staff 124
  staff management 126
Stages of development 6
Standard deviation 22, 102
Standard error 19
Statistical Office (Central/National) 3, 8, 16, 119, 122, 126, 128
Statistician
  functions of 3, 126
  training 125
Stratification 18, 21, 56, 67, 101, 111
Sri Lanka 16, 31, 45, 87, 98, 120
Subjective methods 3, 6, 12, 37
Sudan 14
Supervisor 93, 103, 123, 124, 127
Surveys 96, 130
  food consumption 67
  pilot 12

Tabulation 93, 122
Taiwan 91
Training 124

United Nations 2, 70

Weights *see index number of production*

Zarkovich 41